LARGE-ANGLE CONVERGENT-BEAM ELECTRON DIFFRACTION (LACBED)

Applications to crystal defects

Monograph
of the
French Society
of Microscopies

LARGE-ANGLE CONVERGENT-BEAM ELECTRON DIFFRACTION (LACBED)

Applications to crystal defects

Jean Paul MORNIROLI
Professor in the Ecole Nationale de Chimie de Lille
Laboratoire de Métallurgie et Génie des Matériaux
UMR CNRS 8517
Université des Sciences et Technologies de Lille

SFµ, Paris

Cover illustration: image processing of a LACBED pattern
from a $\Sigma 25$ coincidence grain boundary.
Silicon specimen. Courtesy of J.L. Maurice

Société Française des Microscopies
Case 243 (UPMC) - Bâtiment C
9 Quai Saint Bernard - 75005 PARIS

PREFACE

Electrons are indeed incredible particles! They are not only responsible for the cohesion and the properties of condensed matter but furthermore, they are easily produced, accelerated and, when aimed at a specimen, they interact strongly with it and are scattered, yielding a wealth of information on its organisation at the atomic scale. As they are charged particles, electric and magnetic fields deflect them and it becomes possible to construct devices that act on an electron beam in the same way that a glass lens acts on light. The consequences of all this are enormous; in particular, electron microscopes can be designed, with which we can observe the diffraction pattern associated with the illuminated specimen, in which case we are in Fourier space and we can also observe a magnified image of the specimen, in direct space.

The technique of convergent-beam electron diffraction has brought enormous benefits to structural studies. By replacing an incident beam that is close to being a plane wave by a beam converging strongly on the specimen, the diffraction spots becomes disks; the diffraction information, more spread out in Fourier space, is hence more complete. Large-angle convergent-beam electron diffraction, a natural development, is the ideal form of the technique for the microscopist. The specimen is moved away from the plane conjugate to the recording plane, in which information about both the image and the diffraction pattern is then seen: the two spaces, direct and reciprocal, are both present on the screen of the microscope. We can easily imagine the resulting attractiveness of this method.

A few years ago, during meetings of the Council of the French Society for Microscopies, I suggested that the Society should publish a series of monographs on microscopy. Each monograph should be written by an active specialist of the technique with a strong pedagogic aspect, in order to interest both beginners and experienced microscopists.

Jean-Paul Morniroli is the first author of this series. He has gone far beyond these modest aims and written a textbook that contains an extremely thorough account of the techniques of large-angle convergent-

i

beam diffraction and explains in detail the various microscope operating modes.

That first monograph in French as well as the present English version will serve as a reference for forthcoming books. Jean-Paul Morniroli is to be congratulated on having set the standard so high.

Richard Portier
ENSC Paris

ACKNOWLEDGEMENTS

I make a point of addressing especial thanks to Paul-Henri ALBAREDE, Jeanne AYACHE, David BIRD, Daniel CAILLARD, David CHERNS, Patrick CORDIER, Brigitte DECAMPS, Nathalie DOUKHAN, Joël FAURE, Denis GRATIAS, Christian JÄGER, Wolfgang JÄGER, Elzbieta JEZIERSKA, Danielle LAUB, Jean-Luc MAURICE, Jean-Pierre MICHEL, Maria Louisa NO, Jaime PONS, Richard PORTIER, Sophie POULAT, Alastair PRESTON, Louisette PRIESTER, Abdelkrim REDJAIMIA, Pedro RODRIGUEZ, Pierre STADELMANN, John STEEDS and Michiyoshi TANAKA. They entrusted me with specimens and furnished several of the documents used to illustrate this monograph.

I also thank very warmly the students involved in the LACBED studies of dislocations and grain boundaries. The contributions of Francis GAILLOT, Agnès LECLERE, Olivier RICHARD, Françoise STRZELCZYK and Philippe VERMAUT were particularly appreciated.

During these last few years, many fruitful cooperations have been established with colleagues from Lille and with microscopists in France and elsewhere. They represent a major contribution to this monograph for they gave me the opportunity of observing a very wide range of specimens and brought me into contact with a variety of research fields connected with structure and microstructure. Their role is warmly acknowledged.

Paul-Henri ALBAREDE deserves particular thanks. He developed the tripod method used to prepare flat specimens with nearly constant thickness by mechanical polishing. Such specimens are perfectly adapted to CBED and LACBED experiments. Many of the high-quality patterns shown in this book were obtained from specimens he prepared for me.

Patrick CORDIER, Paul-Henri ALBAREDE and Richard PORTIER accepted the arduous task of proofreading the original French draft of the manuscript. Their remarks, comments, corrections and suggestions were invaluable.

It is a sad duty to mention the name of Mike STOBBS, deeply regretted by the electron microscopy community. He showed me how to obtain my first CBED pattern. That was in 1986, during a stay in Cambridge where my initial goal was to learn how to pratice high resolution electron microscopy! My predilection for reciprocal space began there.

I am very indebted to John STEEDS, David CHERNS and Roger VINCENT, who have shared their extensive knowledge and experience of CBED and LACBED with me. David CHERNS's contribution to the characterization of dislocations by LACBED was a major step forward and doubtless triggered the development of this technique. I appreciated very much the work we did together on the characterization of partial dislocations and grain boundary dislocations.

Many thanks to John SPENCE and Jin Min ZUO. They welcomed me in their laboratory in Tempe, where I learned a great deal about energy filtering and quantitative electron diffraction.

It was Richard PORTIER who launched the idea that the French Society of Microscopies, of which he was then President, should publish a series of monographs on microscopy. I thank him particularly for having encouraged me to embark on this first monograph, which, I hope, will be followed by numerous others. Richard PORTIER has always supported and encouraged convergent-beam electron diffraction. He was one of the few French pioneers involved in CBED in the early eighties.

Most of the experiments were performed on the CM30 Philips transmission electron microscope of the Centre Commun de Microscopie Electronique (CCME) de l'Université des Sciences et Technologies de Lille. Many thanks to the CCME team for providing excellent facilities for LACBED experiments.

I am very grateful to the Head of the Laboratoire de Métallurgie Physique et Génie des Matériaux, Jacques FOCT, for his confidence and support. Since my arrival in his laboratory in 1989, he has always encouraged me to follow this field of reseach.

The idea of preparing an English version of this monograph arose from demands from microscopists unfamiliar with the French language. It was also a good opportunity to incorporate new and recent developments of the LACBED technique in point defects, antiphase boundaries, trace analysis… But the real point of departure was the proposal by one of my colleagues, Etienne Brès, to translate the French version into English. I would like to express here my very sincere thanks to him for preparing this excellent translation in a relatively short period.

Peter Hawkes very kindly agreed to read through the manuscript a last time. I must say that without his careful and meticulous correction of the text, this English version could not have been completed so rapidly. I greatly appreciated his help in the final stages of publication. He also made me realise that the English language is full of subtleties.

I dedicate this monograph to Céline, Florence and Marie-Annick. They helped me a lot with the English langage and "supported" me in both the French and English meanings of the word, during the writing periods of this monograph.

CONTENT

CONTENT

CONTENT

CONTENT

INTRODUCTION

For several decades, the Selected-Area Electron Diffraction (SAED), recommended by Le Poole in 1947 [i.1], was the only electron diffraction technique effectively used. It was regarded for a long time as the "poor relation" of X-ray and neutron diffraction because of its indifferent performance. The accuracy in the measurement of lattice parameters is indeed poor. In addition, the diffracted intensities are strongly disturbed by dynamical phenomena as well as thickness and orientation variations in the diffracted area. As a result, intensity measurements are practically useless, which means that the determination of the structure factor corresponding to the nature and the positions of the atoms in the unit cell cannot be obtained from these patterns except in rare circumstances.

In spite of these weaknesses, the technique is much used because it has two major advantages over the two other diffraction techniques:
- the analysis can be carried out at a microscopic scale,
- the diffraction pattern and the image of the diffracted specimen area are both available in an electron microscope.

Since the mid-seventies, the Convergent-Beam Electron Diffraction (CBED) technique has been developed under the impulse of two research groups: the Bristol group directed by John Steeds and the Sendai group directed by Mishiyoshi Tanaka.

Convergent-beam electron diffraction is in fact the oldest electron diffraction technique since the first CBED experiments were carried out, as early as 1939, on a mica specimen by Kossel and Möllenstedt [i.2]. Nevertheless, it was necessary to await the advent of analytical electron microscopes, with the ability to form convergent incident beams and to overcome the problems of contamination related to this type of illumination, for the convergent-beam electron diffraction technique to develop. This development is also related to an early original and significant application, namely, the identification of the crystal point group at microscopic and nanoscopic scales (Buxton *et al.* [i.3]).

Since then, other very significant applications have appeared:
- space group identification from the observation of Gjønnes and Moodie lines present inside kinematically forbidden reflections [i.4],
- thickness measurement of thin crystals. This measurement is very useful in microanalysis [i.5],
- accurate measurement of lattice parameters. This is used for characterizing local strains.

Why use a convergent beam instead of a parallel one since the theory of convergent beam electron diffraction is much more complex than that of parallel beam diffraction?
For two major reasons.
- The first reason is connected with the information available in the patterns. With a convergent incident beam, the reflections are no longer spots but disks, inside which features and symmetries can be observed. If these symmetries are taken into account, 31 different types of diffraction patterns are obtained and constitute the 31 "diffraction groups" [i.3]. This is more than the ten possibilities of selected-area diffraction patterns, which correspond to the ten planar point groups.
- The second reason is related to the size of the diffracted area. With the selected-area electron diffraction technique, the smallest size of the diffracted area is about 500 nm. At the scale of a thin crystal, this is very large and results inevitably in the presence of thickness variations and local misorientations of the diffracted lattice planes. As the diffracted intensities are extremely sensitive to these variations, selected-area patterns are thus only "average" patterns. Moreover, electron diffraction is characterized by very strong interactions between the transmitted and the diffracted beams known as "dynamical" interactions, which contribute an additional disturbance to the diffracted intensities. These problems can be partly solved by using a convergent incident beam. With the best current microscopes, the probe size is about a few tenths of a nanometre, so that the diffracted area of the specimen can be regarded as constant both in thickness and in orientation. The diffraction pattern is then typical of the crystal. Of course, this pattern is also affected by dynamical interactions, but these interactions do not modify the pattern symmetries and may even improve their visibility.
The convergence angle is in the range 0.1° to 0.5°. This value may seem small but it becomes very significant if it is compared with the Bragg angles. The latter are of the order of a few tenths of a degree in electron diffraction.
As the diameter of the diffracted disks depends on the convergence angle of the incident beam, it is advisable to increase this angle in order to reveal more information inside the transmitted and

diffracted disks. Unfortunately, as soon as the disks overlap to form patterns known as Kossel patterns, the quality of the patterns in the overlapping areas deteriorates. Several solutions have been proposed to overcome this difficulty and are described in chapter IX. The best method is that proposed by Tanaka in 1980 [i.6], which consists in raising or lowering the specimen with respect to its normal position in the microscope. The method is called LACBED for Large-Angle Convergent-Beam Electron Diffraction or the Tanaka method.

The change of the specimen height in the microscope increases the size of the illuminated area, which can reach several micrometres. At first sight, this appears to be a major drawback that will limit the field of application of this method to the analysis of large unfaulted single crystals of uniform thickness. Moreover, a blurred image of the illuminated area is superimposed on the diffraction pattern. For all these reasons, the LACBED technique could have remained a mere curiosity. In fact, **these limitations prove to be major advantages for some specific applications**. The simultaneous observation of the diffraction pattern and the image means that the patterns contain information on the direct and reciprocal spaces and can be regarded as diffraction-image mappings. These mappings are very sensitive to any local modification of the diffraction conditions, for example, those produced by crystal defects.

In this way, Cherns and Preston [i.7] have shown, in 1986, that dislocations produce typical effects on LACBED patterns. These effects are very easy to interpret since they are directly related to the Burgers vector **b** of the dislocations. This is a considerable advantage over the conventional Burgers vector identification techniques.

The LACBED technique was subsequently applied successfully to the analysis of other crystal defects such as stacking faults and grain boundaries. New applications as well as new variants continue to appear.

The first part of this book deals with the theoretical description of LACBED. Since this technique is rather complex, we start by explaining the simpler case of patterns produced by a parallel incident beam. We consider the case where only a single set of lattice planes (hkl) is at the exact Bragg condition or close to it. These are the "two-beam" conditions familiar to microscopists. We continue by examining the "many-beam" conditions and pay special attention to zone-axis patterns, which are a special case of the many-beam conditions.

We then consider the formation of CBED patterns obtained with a small convergence angle after which we describe the Kossel patterns formed by a highly convergent incident beam where the transmitted and diffracted disks partially overlap.

We have thus presented the elements and notions needed to describe the LACBED technique.

In the second part, the experimental conditions required for obtaining LACBED patterns are described. We have deliberately gone into considerable detail as experimental LACBED patterns are easy to obtain provided that the various components of the microscope are correctly adjusted.

We also describe the LACBED variants and pay particular attention to the CBIM (Convergent-Beam Imaging) and the "defocus" CBED, which both possess useful advantages for the analysis of the crystal defects.

The last part concerns the applications of LACBED. Some of these are not original and simply bring some improvements to CBED by virtue of the large convergence angle. We only give the references to these applications. Other applications are original and are made possible only through the use of LACBED. These are primarily applications that use the diffraction-image mapping obtained by defocusing the illumination of the specimen. The applications concerning the study of the crystal defects are explained in detail since LACBED proves to be most useful in this field.

This introduction would be incomplete if it made no mention of precursors such as Jacques Beauvillain, who to our knowledge was the first to describe the LACBED configuration, as early as 1970 [i.8]. We also mention the work of D.J. Smith and J.M. Cowley [i.9] who used defocused Kossel patterns in 1971.

CHAPTER I
Bragg's law

I.1 - Analogy with the reflection of visible light

The diffraction of radiations (X-rays, neutrons and electrons) by crystals results from constructive and destructive interferences of waves. The geometrical aspect of diffraction is explained in a very simple way when considered as a **selective reflection, which presents a strong analogy with the reflection of visible light** by a plane mirror. This last phenomenon satisfies the law of reflection (Figure I.1a):

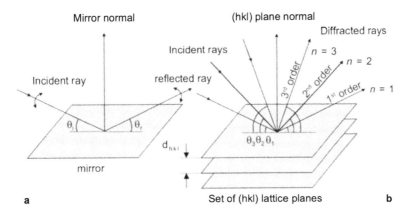

Figure I.1 - Analogy of diffraction with the reflection of visible light by a plane mirror.
a - Reflection of visible light: a reflected ray is obtained whatever is the angle of incidence θ_i of the incident ray with respect to the mirror.
b - Diffraction: a few diffracted rays are only obtained for some well-defined incident angles $\theta_1, \theta_2...\theta_n$ for which Bragg's law $n\lambda = 2d_{hkl}\sin\theta_n$ is satisfied.
In both cases, the reflection or the diffraction angle is equal to the incident angle. The incident ray, the reflected or diffracted rays and the normal to the mirror or to the set of (hkl) lattice planes are located in the same plane.

a - it occurs for any incidence angle θ_i of the incident ray with respect to the mirror,

b - the reflection angle θ_r is always equal to the incidence angle θ_i.

c - the incident ray, the reflected ray and the normal to the mirror lie in the same plane. This plane is perpendicular to the mirror.

Note
According to the conventions commonly used in crystallography, the angles of incidence and reflection θ_i and θ_r are defined with respect to the mirror plane, and not to the mirror normal, as is usual in geometrical optics.

In the case of diffraction (Figure I.1b), the mirror is replaced by a set of (hkl) lattice planes of the crystal with interplanar spacing d_{hkl}.

Conditions (b) and (c) remain unchanged. On the other hand, condition (a) is no longer satisfied, since only a few diffracted rays are observed for well-defined angles of incidence θ_1, θ_2, θ_3...θ_n.

What are these angles?

They satisfy Bragg's law: $n\lambda = 2d_{hkl}\sin\theta_n$ (I.1)

n is the interference order (a positive, usually small integer),
λ is the wavelength of the incident beam,
θ_n is the Bragg angle.

Thus, a set of (hkl) lattice planes gives n diffracted rays. A first-order hkl diffracted ray is obtained for n = 1, a second-order ray for n = 2 and so on (Figure I.1b).

To simplify, only the first-order diffracted rays are taken into account. The others, with orders n > 1, are regarded as first-order reflections on lattice planes with indices (nh nk nl) and with interplanar spacing d_{hkl}/n (Figure I.2b). In this way, a single hkl diffracted ray is assigned to each set of (hkl) lattice planes and Bragg's law is simply written: $\lambda = 2d_{hkl}\sin\theta_B$ (I.2)

Note
Strictly speaking, the (nh nk nl) planes are not lattice planes since only 1/n planes contain direct lattice nodes. For this reason, they are sometimes called reflecting planes.

I.2 - Three-dimensional description of Bragg's law

Figures I.1 and I.2 are useful to explain the formation of electron diffraction patterns obtained with an incident parallel beam because all the incident electrons have the same orientation. Nevertheless, they are not well adapted for the description of electron diffraction patterns obtained with a convergent incident beam. In this case, electrons having different orientations can satisfy Bragg's law simultaneously to produce diffracted beams.

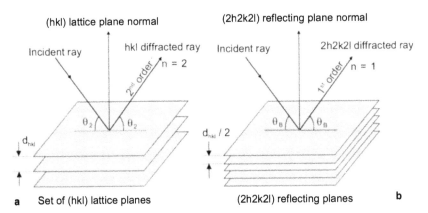

Figure I.2 - Simplification of Bragg's law: a single hkl diffracted ray is assigned to a set of (hkl) lattice planes.
a - Example of the second-order hkl diffracted ray produced by the set of (hkl) lattice planes.
b - The same diffracted ray is considered as a first-order diffracted ray on the (2h 2k 2l) reflecting planes with interplanar spacing $d_{hkl}/2$.

Thus, for each set of (hkl) lattice planes, it is useful to consider the geometrical locus where the incident rays satisfy Bragg's law. As shown on figure I.3a, this is a cone (often called **Kossel cone**) located around the lattice plane normal, and with vertex semi-angle $\pi/2 - \theta_B$.

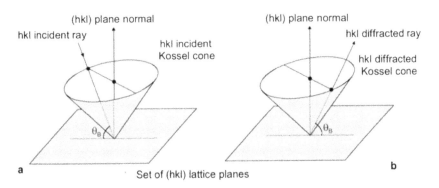

Figure I.3 - Geometrical loci where Bragg's law is satisfied for a set of (hkl) lattice planes.
a - For incident rays, this is a cone with vertex semi-angle $\pi/2 - \theta_B$ and with axis perpendicular to the (hkl) lattice planes.
b - For diffracted beams, it is the same cone as the incident cone.

A diffracted cone corresponds to this incident cone (Figure I.3b). Since the Bragg angle θ_B is identical with the incidence angle θ_i, the incident and diffracted cones are also identical.

The diffraction phenomenon can also occur on the other side of the (hkl) planes, i.e. for the set of ($\bar{h}\,\bar{k}l$) lattice planes corresponding to the $\bar{h}\,\bar{k}l$ incident and diffracted Kossel cones shown on Figure I.4.

hkl lattice planes

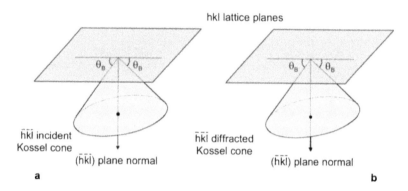

$\bar{h}\,\bar{k}l$ incident Kossel cone

($\bar{h}\bar{k}l$) plane normal

$\bar{h}\,\bar{k}l$ diffracted Kossel cone

($\bar{h}\bar{k}l$) plane normal

a

b

Figure I.4 - Geometrical loci where Bragg's law is obeyed for the set of ($\bar{h}\,\bar{k}l$) lattice planes.
a - For the incident rays, this is a cone with vertex semi-angle $\pi/2$ - θ_B and with axis perpendicular to the ($\bar{h}\,\bar{k}l$) lattice planes.
b - For the diffracted rays, it is the same cone as the incident one.

I.3 - The particular case of electron diffraction

Electron diffraction has two particularities with respect to Bragg's. law.

The first particularity is relevant to the specimens examined in transmission electron microscopy. Usually, they are thin crystals. This means that, in addition to the incident and diffracted rays, undeviated transmitted rays are also observed. The geometrical locus of these transmitted rays is also a cone with vertex semi-angle $\pi/2$ - θ_B (Figure I.5). This cone is common to the $\bar{h}\,\bar{k}l$ incident and diffracted Kossel cones. We note that the angle between a diffracted ray and its corresponding transmitted ray is $2\theta_B$ (called the diffraction angle).

The second particularity is connected with the wavelength of the fast incident electrons. As this wavelength is very small, the Bragg angles are also very small and Bragg's law can be simply written:

$$\lambda \approx 2d_{hkl}\theta_B \qquad (I.3)$$

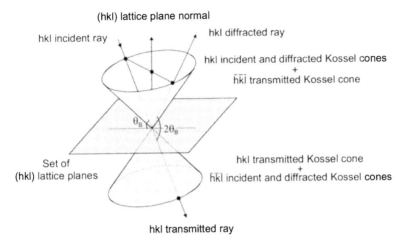

Figure I.5 - Incident, diffracted and transmitted Kossel cones for a set of (hkl) lattice planes.

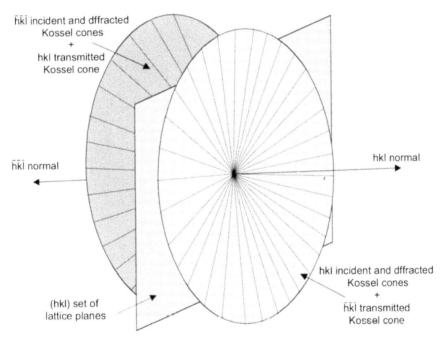

Figure I.6 - Incident, diffracted and transmitted Kossel cones in the case of electron diffraction by a set of (hkl) lattice planes. Owing to the small value of the Bragg angles θ_B, these cones are very flat. The set of (hkl) lattice planes is drawn vertically in order to describe their actual disposition in a transmission electron microscope.

For example:
- the wavelength of electrons accelerated at 100 kV is $\lambda = 0.0036$ nm,
- the corresponding Bragg angle θ_B for the silicon (220) lattice planes is only 0.55°.

A consequence of this very small value of the Bragg angles is that the incident, diffracted and transmitted cones are very flat and have rather the aspect given on Figure I.6. As opposed to the previous figures, the set of (hkl) lattice planes is now drawn vertically. We choose this arrangement because it is closer to the real disposition of lattice planes in an electron microscope. Indeed, in such a microscope:

- the incident electron beam is almost always directed vertically,
- the lattice planes likely to diffract electrons are the planes parallel, or almost parallel, to the incident beam. They are thus vertical or almost vertical.

CHAPTER II
Formation of the diffraction pattern in the electron microscope

Electron diffraction patterns are usually produced with transmission electron microscopes. These instruments are composed of several magnetic lenses. The main lens is the objective lens, which, in addition to forming the first magnified image of the specimen, also produces the first diffraction pattern. This original diffraction pattern is then magnified by the other lenses of the microscope so as to produce the final diffraction pattern on the microscope screen. In order to study the formation of a diffraction pattern in an electron microscope, we just need to consider the electron ray-paths at the level of the objective lens.

II.1 - Electron ray-paths in the objective lens

The objective lens of a transmission electron microscope is a magnetic lens, which can be compared in a first approximation to a thin convergent optical lens. In this chapter, we assume that this lens is ideal, which means we can consider the Gaussian conditions and apply the laws of geometrical or paraxial optics (the rays remain close to the optic axis both in terms of angle and distance).

Note

Large-angle convergent-beam patterns are formed with electrons that travel far from the optic axis. Owing to the spherical aberration of the magnetic lenses, the above ideal conditions are no longer valid. These special features will be described in chapter VIII.

Under these assumptions, the objective lens (Figure II.1a) is characterized by:
- an optic axis,
- an optical centre O,
- front F' and back F focal points located on the optic axis,
- a focal length f = OF = OF',
- front and back focal planes perpendicular to the optic axis at F and F'.

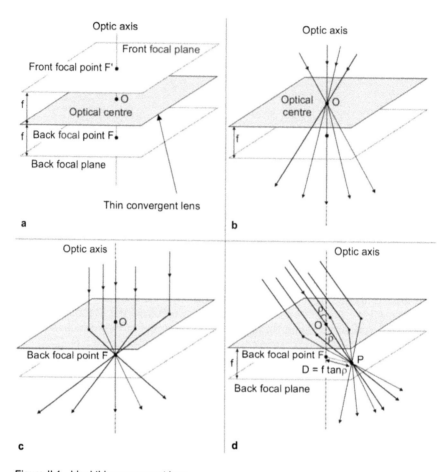

Figure II.1 - Ideal thin convergent lens.
a - Description of the lens characteristics.
b - The rays through the optic axis O are not deviated.
c - The rays parallel to the optic axis are focused at the back focal point F.
d - The rays tilted at an angle ρ with respect to the optical axis are focused at a point P in the back focal plane situated at a distance $D = f\tan\rho$ from the optic axis.

A thin convergent lens has three significant properties:

a - the incident rays passing through the optical centre O are not deviated (Figure II.lb),

b - the incident rays parallel to the optic axis are focused at the back focal point F (Figure II.1c),

c - the incident rays tilted at an angle ρ with respect to the optic axis are focused at a point P in the back focal plane (Figure II.1d). This point is at a distance $D = FP = f \tan\rho$ from the back focal point F.

The electron ray-paths for all the diagrams given in this book can be drawn thanks to these three properties.

Thus, in order to obtain the image A'B' of an object AB located in a plane perpendicular to the optic axis (the object plane), we just need to consider (Figure II.2):

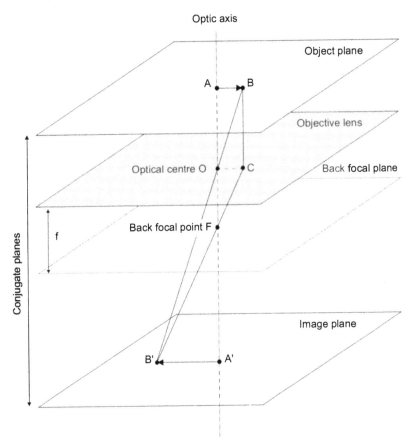

Figure II.2 - Electron ray-paths at the level of the objective lens of an electron microscope. The image A'B' of an object AB located in the object plane is obtained by considering the ray BO, which passes through the optical centre O, and the ray BC, parallel to the optic axis. The first ray is not deviated and the second one passes through the back focal point F. The object and the image planes are conjugate.

- the undeviated ray BO, which passes through the optical centre O of the lens,

- the ray BC, which is parallel to the optic axis and passes through the back focal point F.

This image is inverted with respect to the object itself; it is located in the image plane, conjugate to the object plane. Its magnification γ is:

$$\gamma = A'B'/AB \qquad (II.1)$$

CHAPTER III
Electron diffraction patterns produced by a parallel incident beam

In order to understand how a large-angle electron diffraction pattern is formed by the objective lens of the microscope, it is useful to describe first the formation of a diffraction pattern in the much simpler case of a parallel incident beam.

These experimental conditions correspond to those encountered in "Selected-Area Electron Diffraction" (SAED) (Figure III.1a). We can also consider that they correspond to the Riecke (Figure III.1b) or to the microdiffraction techniques (Figure III.1c) in which a pattern is formed with a parallel or nearly parallel incident beam having a very small spot size.

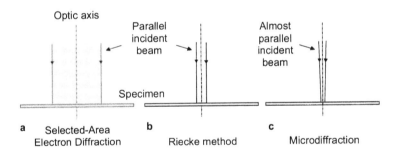

Figure III.1 - Electron diffraction with a parallel or almost parallel incident beam. Illumination conditions of the specimen.
a - Selected-Area Electron Diffraction (SAED). The incident electron beam is parallel and has a diameter of a few microns.
b - Riecke method. The incident beam is parallel and has a diameter of a few tens of nanometres.
c - Microdiffraction. The incident beam is almost parallel and has a diameter from a few nanometres to a few tens of nanometres.

At first, we consider the case where a single set of (hkl) lattice planes produces a diffracted beam. These conditions correspond to the so-called "**two-beam**" conditions (the diffraction pattern contains only the transmitted beam and a single diffracted beam).

Let us describe in detail these two-beam conditions for several reasons:
- they are frequently used in the TEM characterization of crystal defects,
- they are described by relatively simple and well-established equations,
- they are most often found on large-angle convergent-beam electron diffraction patterns.

We will then extend the analysis to several sets of (hkl) lattice planes, i.e. to the "**many-beam**" conditions.

III.1 - Diffraction pattern in the two-beam conditions

Two cases are to be considered:
- a set of (hkl) lattice planes is **exactly** at the Bragg orientation with respect to the incident beam,
- a set of (hkl) lattice planes is **close to** the Bragg orientation.

III.1.1 - A set of (hkl) lattice planes is exactly at the Bragg orientation: exact two-beam conditions

III.1.1.1 - Formation of the diffraction pattern

We consider figure III.2. It shows:
- a parallel incident electron beam directed along the optic axis of the microscope. The diameter of this beam depends on the experimental conditions. It is roughly a few microns in diameter for selected-area electron diffraction.
- a thin single-crystal specimen, which is a slab with parallel top and bottom faces. We suppose that it is oriented so that only one of its sets of (hkl) lattice planes is exactly at the Bragg orientation with respect to the incident beam (Bragg's law is obeyed since the angle between this (hkl) set and the incident beam is equal to the Bragg angle θ_B). At the exit face of the specimen we observe:
- a transmitted beam with the same direction as the incident beam,

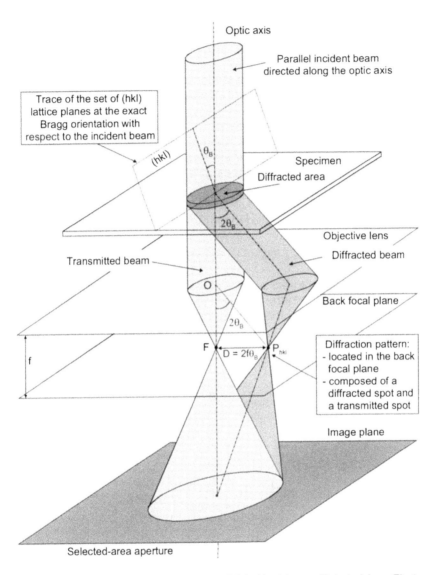

Optic axis

Parallel incident beam
directed along the optic axis

Trace of the set of (hkl)
lattice planes at the exact
Bragg orientation with
respect to the incident beam

(hkl)

θ_B

Specimen

Diffracted area

$2\theta_B$

Objective lens

Transmitted beam

Diffracted beam

O

Back focal plane

$2\theta_B$

F $D = 2f\theta_B$ P_{hkl}

f

Diffraction pattern:
- located in the back
  focal plane
- composed of a
  diffracted spot and
  a transmitted spot

Image plane

Selected-area aperture

Figure III.2 - Electron diffraction with a parallel incident beam (Selected-Area Electron Diffraction).
Electron ray-path at the level of the objective lens when a single set of (hkl) lattice planes is at the exact Bragg orientation with respect to the incident beam directed along the optic axis of the microscope. The diffraction pattern is composed of a transmitted spot and of a single diffracted spot. It is located in the back focal plane of the objective lens.
The diffracted area is selected with the selected-area aperture located in the image plane of the objective lens.

- a diffracted beam deviated by an angle $2\theta_B$ (diffraction angle) with respect to the transmitted beam.

In agreement with the properties of a thin convergent lens described in the previous chapter:

- all the transmitted rays are focused at the back focal point F,
- all the diffracted rays are focused at a point P_{hkl} located in the back focal plane.

Then, the diffraction pattern:

- **is located in the back focal plane of the objective lens,**
- **is composed of a transmitted spot and of a diffracted spot. For this reason, it is described as a "spot pattern".**

In the back focal plane, the distance D between the diffracted and the transmitted spots is related to the Bragg angle θ_B by the formula:

$$D = f\tan 2\theta_B \qquad\qquad (III.1)$$

Since the Bragg angles θ_B are very small in electron diffraction

$$D \approx 2f\theta_B \qquad\qquad (III.2)$$

The other magnetic lenses of the microscope magnify this original diffraction pattern in order to form the final diffraction pattern on the microscope screen. On this screen, the distance D becomes D_{hkl} (Figure III.3) with:

$$D_{hkl} \approx 2L\theta_B \qquad\qquad (III.3)$$

L is the **camera length**.

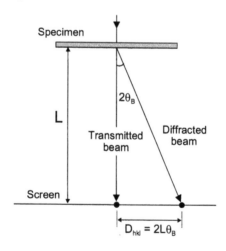

Figure III.3 - Formation of the diffraction pattern on the screen of the microscope. The distance D_{hkl} between the transmitted beam and the diffracted beam is directly related to the Bragg angle θ_B and to the camera length L.

According to Bragg's law (I.3), $\theta_B = \lambda/2d_{hkl}$ (III.4)

hence:

$D_{hkl} = L\lambda/d_{hkl}$ (III.5)

$D_{hkl}d_{hkl} = L\lambda = C$ (III.6)

The term $L\lambda$ depends only on the experimental conditions. It is called the **camera constant C**, and is very useful for indexing diffraction patterns. An example of pattern interpretation is given in paragraph X.4.1.1.1.

Note
In selected-area electron diffraction, the diffracted area of the specimen is selected with the selected-area aperture located in the image plane of the objective lens. As shown on figure III.2, this aperture limits the diffracted area. On this figure, only the beams, which pass through this aperture, are shown. In practice, the parallel incident beam is usually much larger than the diffracted area. The selected-area aperture could also be placed near the object plane where it would have the same effect. However, it is more convenient to place it in the image plane where it acts on an already magnified image and where more space is available.

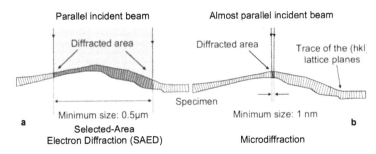

Figure III.4 - Nature of the diffracted area.
a - Selected-area electron diffraction. The diffracted area is large and contains significant thickness changes and orientation variations of the (hkl) lattice planes. The resulting diffraction pattern is "averaged".
b - Microdiffraction. The diffracted area can be very small so that the thickness and the orientation of the lattice planes in the diffracted area can be regarded as constant. Thus, the diffraction pattern is typical of the crystal.

As a result of the spherical aberration of the objective lens, the minimum size of the diffracted area is about 500 nm. Consequently, it inevitably contains thickness variations as well as orientation variations of the lattice planes (Figure III.4a). The selected-area electron diffraction pattern is then an average diffraction pattern. This important limitation no longer exists with the Riecke or the microdiffraction techniques where the diffracted area is very small since it is directly defined by the size of the incident beam. The minimum size of the incident beam is about 1 nm with the best recent microscopes. Under these experimental conditions, we can consider that the thickness and the orientation of the lattice planes in the diffracted area are constant (Figure III.4b). The diffraction pattern is then typical of the crystal under examination.

III.1.1.1 - Ewald sphere construction

The Ewald sphere construction is a very simple way of describing the diffraction phenomenon. It takes into account both the direct lattice (the crystal) and its reciprocal lattice (the diffraction pattern).
This construction consists of (Figure III.5a):
- a sphere of radius $R = 1/\lambda$, which characterizes the wavelength λ of the incident beam,
- the incident beam directed along the radius AO,
- the specimen located at the centre O of the Ewald sphere,
- the reciprocal lattice, which is drawn with its origin O* located at the end of the diameter AO*.

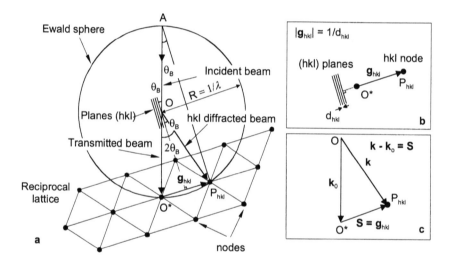

Figure III.5 - Ewald sphere construction. General case.
a - A diffracted beam is produced if an hkl node of the reciprocal lattice is located exactly on the Ewald sphere. This situation occurs for the node P_{hkl}.
b - Relationship between a set of (hkl) lattice planes and its corresponding reciprocal lattice vector g_{hkl}. This vector is perpendicular to the (hkl) lattice planes and its modulus g_{hkl} is equal to $1/d_{hkl}$.
c - Vector description of Bragg's law. Bragg's law is satisfied when the scattering vector S is equal to the reciprocal lattice vector g_{hkl}.

We recall that each set of (hkl) lattice planes of the direct lattice is characterized by a g_{hkl} vector of the reciprocal lattice (Figure III.5b) with:
$$g_{hkl} = ha^* + kb^* + lc^* \qquad (III.7)$$
a^*, b^* and c^* are the basis vectors of the reciprocal lattice. This vector is perpendicular to the (hkl) set, and $g_{hkl} = 1/d_{hkl}$ $\qquad (III.8)$

A node hkl of the reciprocal lattice is located at the end of this vector. The nodes of the reciprocal lattice form rows and planes; the latter are called reciprocal layers.

An hkl diffracted beam is produced when an hkl node of the reciprocal lattice is exactly located on the Ewald sphere. In this case, Bragg's law is satisfied for the corresponding set of (hkl) lattice planes since:

$$\sin\theta_B = O^*P_{hkl}/AO^* = \lambda g_{hkl}/2 = \lambda/2d_{hkl} \tag{III.9}$$

which corresponds to Bragg's law $\lambda = 2 d_{hkl} \sin\theta_B$

This situation occurs for the hkl node located at point P_{hkl} on figure III.5a. The hkl diffracted beam is directed along OP_{hkl} and makes a diffraction angle $2\theta_B$ with respect to the transmitted beam. The orientation of the (hkl) lattice planes is also given by the Ewald sphere construction.

If $\mathbf{k_0}$ (with modulus $k_O = 1/\lambda$) and $\mathbf{k}$ (with modulus $k = 1/\lambda$) are the wave vectors used to describe both the direction and the wavelength of the incident and diffracted beams, then the Ewald sphere construction (Figure III.5c) shows that Bragg's law can also be written in the form:

$$\mathbf{k} - \mathbf{k_O} = \mathbf{S} = \mathbf{g}_{hkl} \tag{III.10}$$

Note
The diffraction phenomenon results only from elastic scattering. This means that the incident electrons involved in the diffraction phenomenon interact with the crystal without energy loss and therefore without any wavelength change. For that reason, the moduli k_0 and k have the same value.

S is **the scattering vector**. From this point of view, the hkl diffraction phenomenon occurs when the scattering vector **S** is equal to a $\mathbf{g}_{hkl}$ vector of the reciprocal lattice: $\mathbf{S} = \mathbf{g}_{hkl}$ $\hfill$ (III.11)

We can thus express the exact Bragg orientation for a set of (hkl) lattice planes in three different ways:

- Bragg's law $\lambda = 2d_{hkl}\sin\theta_B$ is satisfied,
- the corresponding hkl node of the reciprocal lattice is exactly located on the Ewald sphere,
- the reciprocal lattice vector $\mathbf{g}_{hkl}$ is equal to the scattering vector **S**.

III.1.1.2.1 - Relationship between the Ewald sphere construction and the diffraction pattern

Since the Ewald sphere construction gives the direction of the diffracted beam, the relationship between this construction and the diffraction pattern located in the back focal plane of the objective lens (or on the microscope screen) is obvious. It is shown on figure III.6.

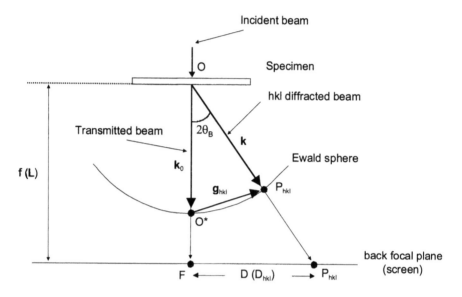

Figure III.6 - Relationship between the Ewald sphere construction and the diffraction pattern located in the back focal plane (or on the microscope screen).

III.1.1.2.2 - Peculiarities of the electron diffraction phenomenon

For a random crystal orientation with respect to the incident beam, the likehood of obtaining a diffracted beam is very small since it is related to the probability that a reciprocal lattice node is present on the Ewald sphere. In order to obtain a diffracted beam, the crystal must be tilted so that an hkl node lies exactly on the Ewald sphere. Actually, this strict condition is only valid for X-ray or neutron diffraction. For electron diffraction, it is less severe owing to the two following peculiarities.

The first peculiarity is related to the very small wavelength λ of the electron beam; it makes the radius $R = 1/\lambda$ of the Ewald sphere very large compared to the moduli g_{hkl} of the reciprocal lattice vectors.

For example, if we consider the silicon 220 reflection produced by electrons accelerated at 100 kV:
- the radius of the Ewald sphere is $R = 1/0.0036 = 277 \ nm^{-1}$
- the modulus of the reciprocal vector is $g_{220} = 1/d_{220} = 1/0.192 = 5.20 \ nm^{-1}$
Thus, R is approximately 50 times larger than g_{hkl}.

Consequently, **the Ewald sphere has a very large radius** and the Ewald sphere construction for electron diffraction looks like the diagram shown on figure III.7a.

The second peculiarity of electron diffraction relates to the nature of specimens. Usually, specimens for electron microscopy are thin foils having a very small thickness t. As a result, a relaxation of **the exact Bragg conditions** occurs:
- the nodes of the reciprocal lattice are no longer spots but are extended along the direction of least specimen thickness. They are thus transformed into short rods known as "relrods" as illustrated on figure III.7a. According to the nature of the (hkl) diffracted lattice planes and to the specimen thickness, the length of the relrods depends on 2/t or on $2/\xi_g$. The quantity ξ_g is the **extinction distance** of the hkl reflection (this distance will be described in chapter VII). The diffracted intensity along a relrod can be as shown on figure III.7a or may be more complex as shown on figure III.7b. In most cases, the maximum diffracted intensity is obtained when reciprocal node is located in the middle of the relrod. This feature will be described in detail in chapter VII.

As a result, we note that a diffracted beam can be produced even if the set of (hkl) lattice planes is not at the exact Bragg orientation. The deviation from the exact Bragg orientation is characterized by:

$$\Delta\theta = \theta_i - \theta_B \qquad\qquad\qquad (III.12)$$

θ_i being the angle of incidence of the electrons with respect to the set of (hkl) lattice planes.

III.1.2 - A set of (hkl) lattice planes is close to the Bragg orientation: near two-beam conditions.

III.1.2.1 - Formation of the pattern

If the set of (hkl) lattice planes is not exactly at the Bragg orientation with respect to the parallel incident beam directed along the optic axis (for example, when the incidence angle θ_i is smaller than θ_B as shown on figure III.8), the two peculiarities of electron diffraction described above imply that a diffracted beam is observed along an angle close to the diffraction angle $2\theta_B$.
This interesting property can be understood with the Ewald sphere construction.

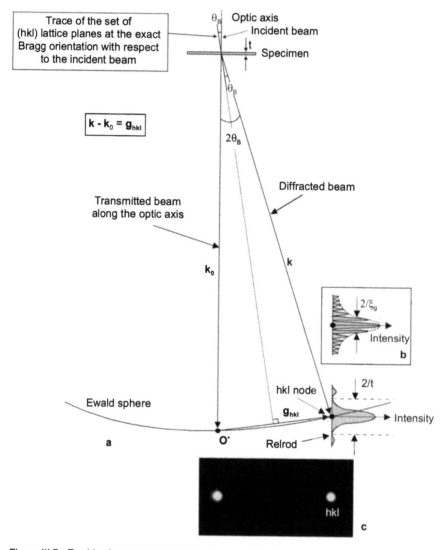

Figure III.7 - Ewald sphere construction for electron diffraction.
a - Ewald sphere construction for a set of (hkl) lattice planes at the exact Bragg orientation with respect to the incident beam directed along the optic axis. The radius of the Ewald sphere is much larger than the modulus of the g_{hkl} vector. The hkl node is located exactly on the Ewald sphere.
b - Complex aspect of the diffracted intensity along a relrod.
c - Example of an experimental pattern under exact two-beam conditions.

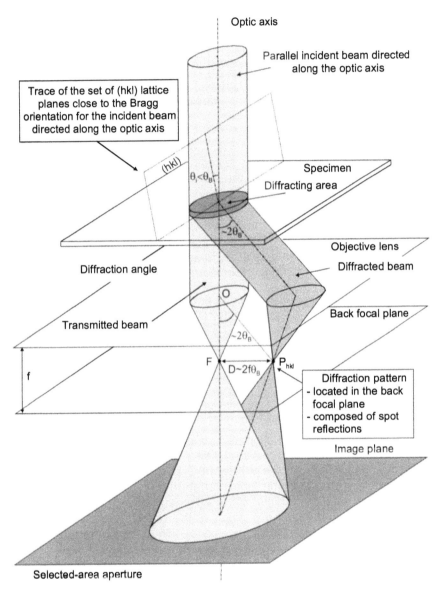

Optic axis

Parallel incident beam directed along the optic axis

Trace of the set of (hkl) lattice planes close to the Bragg orientation for the incident beam directed along the optic axis

(hkl)

$\theta_i < \theta_B$

Specimen

Diffracting area

~2θ_B

Objective lens

Diffraction angle

Diffracted beam

Transmitted beam

O

Back focal plane

~2θ_B

f

F

D~2fθ_B

P_hkl

Diffraction pattern
- located in the back focal plane
- composed of spot reflections

Image plane

Selected-area aperture

Figure III.8 - Electron diffraction with a parallel incident beam.
Electron ray-path at the level of the objective lens when a single set of (hkl) lattice planes is close to the Bragg orientation with respect to the incident beam directed along the optic axis of the microscope. The diffraction angle is very close to 2θ_B. The geometry of the diffraction pattern is not modified by a slight misorientation of the set of lattice planes.

III.1.2.2 - Ewald sphere construction

We consider the Ewald sphere construction (Figure III.9a) corresponding to figure III.8. The hkl reciprocal node is now located at some distance from the Ewald sphere. This distance is characterized by a vector **s** called the **excitation** or **deviation vector**. This is a vector of the reciprocal lattice with:

$$\mathbf{s} = s_x\mathbf{a}^* + s_y\mathbf{b}^* + s_z\mathbf{c}^* \qquad \text{(III.13)}$$

It is parallel to the incident beam $\mathbf{k_o}$ and connects the hkl reciprocal node to the Ewald sphere. By convention, the deviation parameter s is positive if **s** has the same direction as $\mathbf{k_o}$: the node of the reciprocal lattice is then located inside the Ewald sphere and $\Delta\theta$ is positive (Figure III.9c). On the other hand, s is negative if **s** is directed along the opposite direction. In this case, the node is located outside the Ewald sphere as shown on figure III.9a and $\Delta\theta$ is negative. The value s = 0 corresponds to a node exactly located on the Ewald sphere (Figure III.9b), i.e. at the exact Bragg orientation.

We said previously that the node of the reciprocal lattice is actually a relrod. If this relrod is long enough (as it will be for a very thin specimen) and if the node is not too far from the Ewald sphere (at it will be for a small misorientation of the (hkl) lattice planes), it can intersect the Ewald sphere and give a diffracted beam. The large radius of the Ewald sphere greatly favours this operation.

In this case, Bragg's law is rewritten:

$$\mathbf{k} - \mathbf{k_o} = \mathbf{S} = \mathbf{g_{hkl}} + \mathbf{s} \qquad \text{(III.14)}$$

The Ewald construction also states that:
- the diffraction angle is only very slightly affected, which means that **the geometry of the diffraction pattern is not significantly modified by a small crystal misorientation**.
- the intensity of the diffracted beam is very strongly affected because it depends on the position of the relrod, which intersects the Ewald sphere (Figures III.9a, b and c). Thus, **the intensity of the diffraction pattern is very strongly modified by a small crystal misorientation**.

III.1.2.3 - Characterization of the deviation parameter s

The deviation parameter s is an expression of the misorientation $\Delta\theta = \theta_1 - \theta_B$ of the set of (hkl) lattice planes with respect to the exact Bragg orientation.

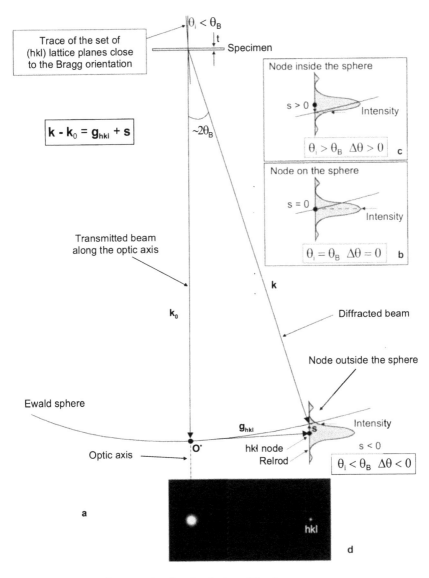

Figure III.9 - Ewald sphere construction for electron diffraction.
a - Ewald sphere construction for a set of (hkl) lattice planes close to the Bragg orientation with respect to the incident beam directed along the optic axis. The hkl node of the reciprocal lattice is located outside the Ewald sphere (s < 0).
b, c - Examples of zero and positive deviation parameter s.
d - Experimental pattern near two-beam conditions. Compare the intensity of the hkl diffracted spot with the one displayed on figure III.7c.

According to figure III.10:

$$s = g_{hkl}\, \Delta\theta \qquad\qquad\qquad\qquad\qquad \text{(III.15)}$$

However,

$$g_{hkl} = 1/d_{hkl} \qquad\qquad\qquad\qquad\qquad\qquad \text{(III.8)}$$

and finally:

$$s = \Delta\theta/d_{hkl} \qquad\qquad\qquad\qquad\qquad\qquad \text{(III.16)}$$

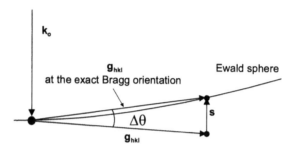

Figure III.10 - Relationship between the excitation vector **s** and the misorientation $\Delta\theta$ of the (hkl) lattice planes with respect to the exact Bragg orientation.

III.1.3 - Special cases

III.1.3.1 - The parallel incident beam is not directed along the optic axis of the microscope

In the previous paragraphs, we made the assumption that the parallel incident beam was directed along the optic axis of the electron microscope. In practice, this ideal configuration is not always found. The incident beam may be tilted, either accidentally when the microscope is misaligned, or on purpose since it is a very convenient way of modifying the orientation of the incident beam with respect to the (hkl) lattice planes. On modern electron microscopes, it is possible to change very accurately the orientation of the incident beam by means of the beam deflection coils, which are normally used to produce the dark-field images. This modification occurs within an angular field that can reach several degrees. This method is an alternative way of tilting the specimen without using the specimen holder of the microscope.

This technique can be used to tilt a set of (hkl) lattice planes to the exact Bragg orientation as shown on figure III.11, where the incident beam is tilted by an angle ρ with respect to the optic axis. Note that the transmitted beam is no longer at the focal point F of the objective lens, but is shifted by a distance $t = f\rho$ $\qquad\qquad\qquad\qquad$ (III.17)

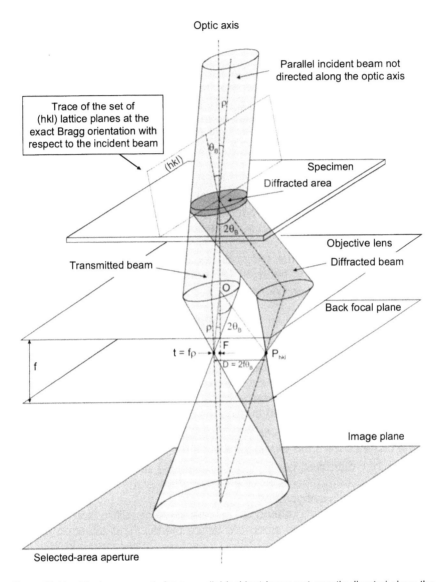

Optic axis

Parallel incident beam not
directed along the optic axis

Trace of the set of
(hkl) lattice planes at the
exact Bragg orientation with
respect to the incident beam

Specimen

Diffracted area

Objective lens

Transmitted beam

Diffracted beam

Back focal plane

$t = f\rho$

$D = 2f\theta_B$

P_{hkl}

f

Image plane

Selected-area aperture

Figure III.11 - Electron ray-path for a parallel incident beam not exactly directed along the optic axis of the microscope.
The incident beam is tilted by an angle ρ with respect to the optic axis, and is thus at the exact Bragg orientation for the set of (hkl) lattice planes. In the back focal plane of the objective lens, the diffraction pattern is shifted by the distance $t = f\rho$ with respect to the optic axis.

The same shift occurs for the diffracted beam since the latter makes an angle $2\theta_B$ with the transmitted beam.

A misorientation of the incident beam with respect to the optic axis leads to a shift of the "whole" diffraction pattern in the back focal plane of the objective lens (and in the conjugate screen of the microscope). The transmitted beam is no longer at the centre of the screen. This property will be used to describe the formation of convergent-beam electron diffraction patterns.

The corresponding Ewald sphere construction is given on figure III.12. The transmitted beam is tilted by an angle ρ with respect to the optic axis. This operation produces a shift (without rotation since the specimen remains stationary) of the reciprocal lattice by a distance:

$$t' = \rho/\lambda \tag{III.18}$$

It also places the hkl node exactly on the Ewald sphere. The experimental pattern of figure III.12b obtained under these conditions is identical with the one on figure III.9b. It is shifted by a distance $t = f\rho$.

III.1.3.2 - The set of (hkl) lattice planes is parallel to the electron beam

A special case is observed when the set of (hkl) lattice planes is parallel to the optic axis. The corresponding reciprocal vector $\mathbf{g}_{hkl}$ is then perpendicular to the optic axis, and the deviation parameter s is negative since the hkl node is located outside the Ewald sphere (Figure III.13).

For this configuration, the opposite $\overline{hkl}$ node is situated at the same distance from the Ewald sphere as the hkl one, which means that a second $\overline{hkl}$ diffracted beam is simultaneously produced. It corresponds to the diffraction on the other side of the lattice planes. Somewhat, this is a **"three-beam"** configuration.

Such a configuration is surprising since diffraction occurs for electrons that propagate along the lattice planes. This situation will also occur for Zone-Axis Patterns (ZAPs), described in paragraph III.2.2.

On the Ewald sphere construction of figure III.13, we have superimposed in dotted lines the ray-paths that correspond to the exact Bragg orientation for the set of (hkl) lattice planes. To reach this orientation, the incident beam must be tilted by an angle θ_B with respect to the optic axis in order to place the hkl node on the Ewald sphere. Since there is an angle of $2\theta_B$ between the transmitted and the diffracted beams, the latter is also tilted by θ_B.

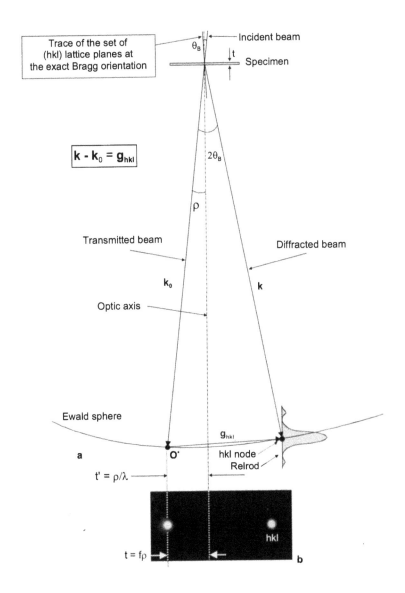

Figure III.12 - Ewald sphere construction when the parallel incident beam is not directed along the optic axis of the microscope.
a - The incident beam is tilted by an angle ρ with respect to the optic axis, in order to place the set of (hkl) lattice planes at the exact Bragg orientation. The reciprocal lattice is shifted by $t' = \rho/\lambda$ so that the hkl node is located on the Ewald sphere.
b - Example of an experimental pattern obtained under these conditions. It is identical with the one shown on figure III.6c except for a shift of $t = f\rho$ with respect to the optic axis.

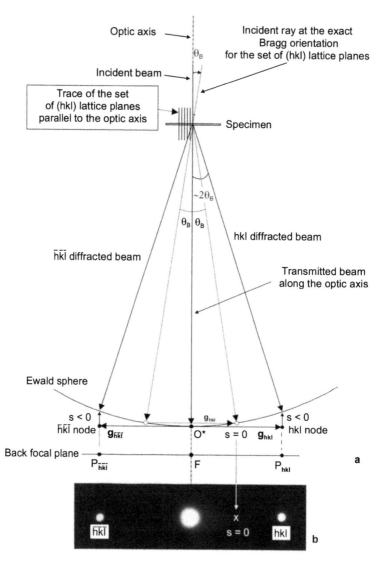

Figure III.13 - Special case of electron diffraction pattern observed when the set of (hkl) lattice planes is parallel to the incident electron beam.
a - Ewald sphere construction. The drawing in dotted lines corresponds to the exact Bragg orientation for the (hkl) set of lattice planes. It indicates that the value s = 0 is located halfway between the points F and P_{hkl}.
b - Example of an experimental pattern obtained under three-beam conditions.

The main interest of this superimposition is to indicate that the value s = 0 of the deviation parameter is located halfway between the transmitted beam F and the diffracted beam P_{hkl}.

III.1.4 - Application: setting (hkl) lattice planes at the exact Bragg orientation

There are two experimental ways of positioning a set of lattice planes (hkl) at the exact Bragg orientation (Figure III.14a).

The first one consists in tilting the specimen by an angle ρ so that the incidence angle θ_i is equal to the Bragg angle θ_B (Figure III.14b). This operation has almost no effect on the position of the hkl reflection but it produces a very significant effect on its intensity (the maximum of intensity being generally obtained for the exact Bragg orientation).

Alternatively, the incident beam may be tilted by an angle ρ in order to place this beam at the exact Bragg orientation (Figure III.14c). The diffraction pattern then obtained is identical with the previous one except that it is shifted from the focal point F by a distance t = fρ.

III.2 - Diffraction pattern in "multi-beam" conditions

III.2.1 - General Case

Until now, we have considered that a single set of (hkl) lattice planes was at the exact Bragg orientation or close to it. Usually, several sets of lattice planes can simultaneously be at the Bragg orientation (exactly or approximately). The description given in the previous paragraphs for a single set of lattice planes then applies to each set involved.

For the Ewald sphere construction, this means that several relrods simultaneously intersect the Ewald sphere (Figure III.15a). The diffraction patterns then display several reflections (Figure III.15b). We note that these multi-beam conditions are all the more favoured since the curvature of the Ewald sphere is small (the acceleration voltage of the electrons is high) and the relaxation of the Bragg conditions is significant (the specimen is thin). Generally speaking, the number of reflections present on the diffraction pattern is related to these conditions.

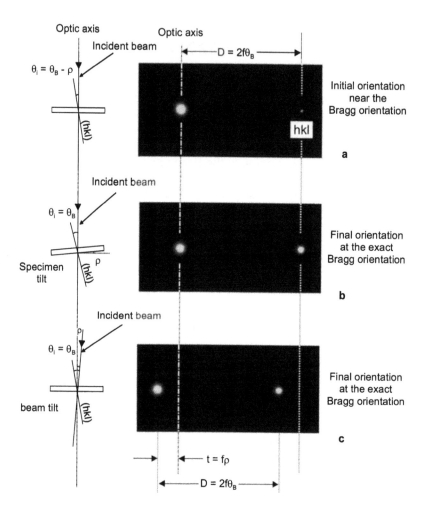

Figure III.14 - Setting (hkl) lattice planes at the exact Bragg orientation from an initial orientation near the Bragg orientation.

a - Initial orientation. The lattice planes are close to the Bragg orientation.

b - The specimen is tilted by an angle ρ until the incidence angle θ_i is equal to the Bragg angle θ_B. During this operation, the intensity of the hkl reflection increases and reaches its maximum value.

c - The incident beam is tilted by an angle ρ until the incidence angle θ_i is equal to the Bragg angle θ_B. The corresponding pattern is identical with the previous one except that it is shifted by the distance $t = f\rho$ with respect to the optic axis.

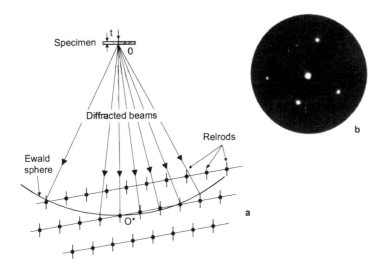

Figure III.15 - "Multi-beam" conditions.
a - Ewald sphere construction. Several relrods simultaneously intersect the Ewald sphere and produce diffracted beams.
b - Example of an experimental diffraction pattern. It is composed of several spot reflections at the exact Bragg orientation or close to it.

III.2.2 - Special case: [uvw] zone axis pattern (ZAP)

A very useful case occurs when a high symmetry [uvw] lattice row of the crystal is set parallel to the incident electron beam. Under these experimental conditions, we can consider that all the hkl nodes of the reciprocal lattice are stacked into parallel and equidistant (uvw)* reciprocal layers. These layers are characterized by an integer index n (Figure III.16a). The hkl nodes located into the n-th layer obey the zone condition $hu + kv + lw = n$ (III.19)

These layers are perpendicular to the [uvw] lattice row of the direct lattice and have an interlayer distance $H = 1/P_{[uvw]}$ (III.20) $P_{[uvw]}$ being the parameter of the [uvw] lattice row, i.e. the distance between two adjacent direct lattice nodes along this row (Figure III.18a).

The Ewald sphere construction on figure III.16a shows that some relrods close to the origin O* and contained in the layer n = 0 intersect the Ewald sphere. They give a set of reflections located in the central area of the pattern. These reflections constitute the **Zero-Order Laue Zone (ZOLZ)** (Figure III.16b). This Laue zone is an enlarged image of the layer n = 0 of the reciprocal lattice.

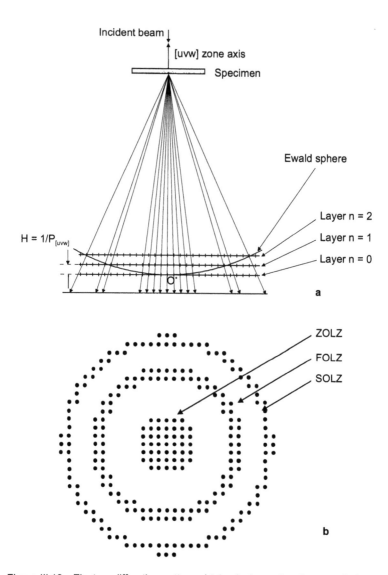

Figure III.16 - Electron diffraction pattern obtained when a [uvw] zone axis is parallel to the incident beam.
a - Ewald sphere construction. The reciprocal lattice nodes (or relrods) are located in parallel and equidistant layers. Some relrods located in the 0th, 1st, 2nd... layers intersect the Ewald sphere and give diffracted beams.
b - Corresponding diffraction pattern composed of Laue zones.

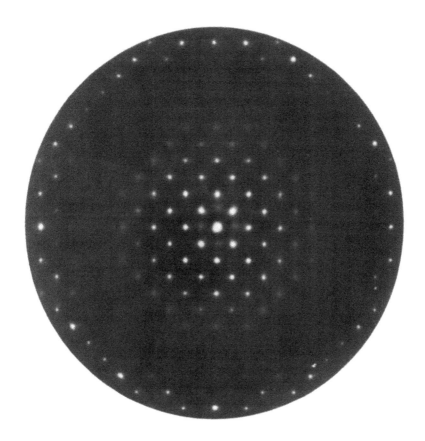

Figure III.17 - Experimental [001] zone axis diffraction pattern from a thin gold foil. The reflections located in the central area of the pattern form the zero-order Laue zone. The reflections of the first-order Laue zone are located on a ring. Only the first-order Laue zone is visible on this pattern.
Gold specimen. Courtesy of J. Ayache.

In the same way, some relrods of the layer n = 1 intersect the Ewald sphere and produce reflections located in a concentric area of the pattern called the **First-Order Laue Zone** (FOLZ). Note that there is no reflection between these two Laue zones. The relrods in the other n layers give reflections located on concentric areas called **High-Order Laue Zones** (HOLZ), which are magnified images of the corresponding n reciprocal layers. An example of an experimental zone axis pattern is given on figure III.17.

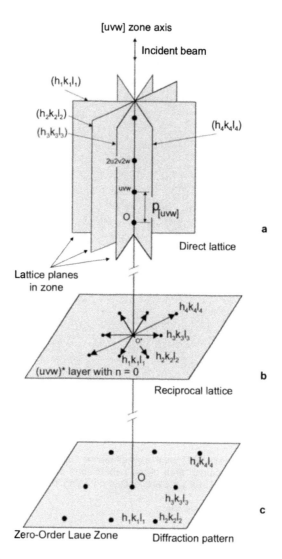

Figure III.18 - [uvw] zone-axis diffraction pattern. The incident electron beam is parallel to the [uvw] zone axis. Relationship between the reflections located in the zero-order Laue zone and the (hkl) lattice planes containing the [uvw] zone.
a - Lattice planes containing the [uvw] zone axis (planes in the zone).
b - (uvw)* layer of the reciprocal lattice with zero order. This layer contains the origin O* of the reciprocal lattice and the hkl nodes that correspond to the planes in the zone.
c - Zero-order Laue zone of the diffraction pattern. It is a magnified image of the zero-order layer of the reciprocal lattice.

These zone-axis patterns display some interesting characteristics:

- the reciprocal layers with negative orders -n are not involved because they cannot intersect the Ewald sphere (some exceptions to this rule can be observed if the incident beam is not exactly parallel to the zone axis or if the specimen is very thin).

- the hkl reflections located in the ZOLZ (Figure III.18c) correspond to hkl reciprocal nodes contained in the zero layer. Since they are situated at the end of g_{hkl} vectors (Figure III.18b), which are perpendicular to the incident electron beam, the corresponding (hkl) lattice planes are parallel to the electron beam as well as to the [uvw] lattice row (Figure III.18a). These planes are said to be **in the zone** with the [uvw] lattice row, which is called the **[uvw] zone axis**. For this reason, the corresponding diffraction pattern is often called a **Zone-Axis Pattern (ZAP)**.

- for each hkl reflection located in the ZOLZ, the special conditions described in paragraph III.1.2 occur since the incident beam is parallel to the lattice planes involved. These planes are then vertical in the electron microscope.

- the ZOLZ only give two-dimensional information about the reciprocal lattice since only one reciprocal layer (with n = 0) is involved.

- the reflections located in the ZOLZ are not exactly in Bragg orientation. This property is illustrated on figure III.19.

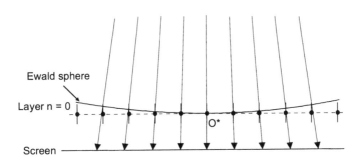

Figure III.19 - Ewald sphere construction in the case of a zone axis pattern.
Enlargement of figure III.16 around the origin O* of the reciprocal lattice showing that the nodes belonging to the zero-order layer are not exactly located on the Ewald sphere. All of them have a negative deviation parameter.

- the HOLZ reflections come from g_{hkl} vectors forming an obtuse angle with the incident beam. They correspond to (hkl) lattice planes tilted with respect to this beam.

Note
The (hkl) lattice planes giving the HOLZ reflections are actually very slightly tilted with respect to the [uvw] lattice row. They are, therefore, almost vertical inside the column of the microscope. In electron diffraction, all the diffracting planes are parallel or almost parallel to the incident beam.

The **Whole Pattern** (WP) composed of the ZOLZ and of all the HOLZs gives **three-dimensional** information about the reciprocal lattice since it is generated by several layers of the reciprocal lattice. This is an extremely useful property.

CHAPTER IV
Diffraction pattern produced by a convergent incident beam: CBED pattern

As in the previous chapter, we start by examining the case of a single set of (hkl) lattice planes in exact or near two-beam conditions.

IV.1 - Diffraction pattern under two-beam conditions

IV.1.1 - A single set of (hkl) lattice planes is at the exact Bragg orientation: exact two-beam conditions

IV.1.1.1 - Formation of the diffraction pattern

We consider the same experimental conditions as those on figure III.2 except the incident beam that is now convergent with a semi-angle α of about 0.5°. For the sake of clarity, this angle is strongly exaggerated on figure IV.1.

> **Note**
> At first sight, the semi-angle value α seems very small. The significant point is that this angle has the same order of magnitude as Bragg angles θ_B.

We also suppose that the axis of this incident convergent beam is directed along the optic axis of the microscope. The diffracted area **is now directly defined by the size of the incident beam** and not by the selected-area aperture as for selected-area electron diffraction (figure IV.2). It is usually in the range 1 to 50 nm so that very small particles can be explored. It is also possible to avoid any thickness changes and variations of the lattice plane orientation in the diffracted area of the specimen. The convergent incident beam can be regarded as composed of a set of **incoherent incident beams** (the incoherent aspect of the beam will be described in chapter VII) **having all the orientations within the convergent beam**.

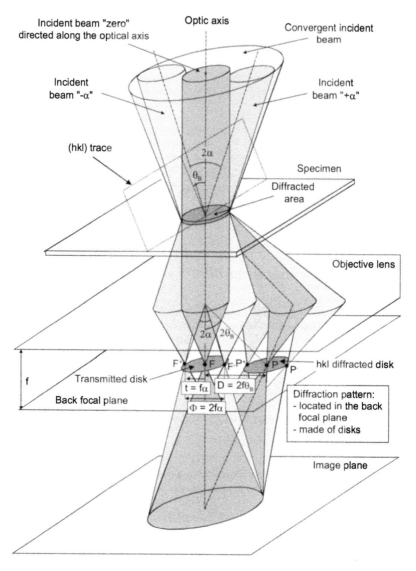

Figure IV.1 - Formation of the diffraction pattern produced by a convergent incident beam. Exact two-beam conditions.

The convergent incident beam is regarded as composed of a set of incident beams having all possible orientations within the incident cone. Three incident beams "zero", "-α" and "+α" are shown on this figure. The incident beam "zero" directed along the optic axis is at the exact Bragg orientation for the (hkl) lattice planes. The diffraction pattern is located in the back focal plane of the objective lens. It is made of an hkl diffracted disk and a transmitted disk.

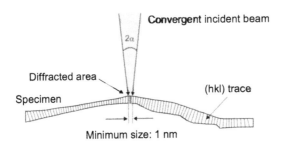

Figure IV.2 - Specimen illumination by a convergent incident beam. The diffracted area is defined directly by the size of the incident beam. The minimum size is about 1 nm with modern microscopes. The diffracted area does not include any thickness change and (hkl) lattice plane variation.

These beams correspond to plane waves whose wave vectors $\mathbf{k_0}$ have different orientations. For the sake of clarity, only three of these incident beams are drawn on figure IV.1:

- the incident beam "zero" directed along the optic axis. It behaves like the incident beam described in paragraph III.1.1.1 and on figure III.2. It produces a transmitted point located at the back focal point F, and a diffracted point P located in the back focal plane at the distance $D = 2f\theta_B$ from the focal point F.

- the incident beam with the extreme orientation "$+\alpha$". This beam behaves like the tilted incident beam described in paragraph III.1.3.1 and on figure III.11. It produces a diffracted point P^+ and a transmitted point F^+. These two points are separated by $D = 2f\theta_B$ and are shifted, with respect to the two previous points F and P, by the distance $t = f\alpha$.

- the incident beam with opposite extreme orientation "$-\alpha$". In the same way, it produces a pair of points P^- and F^- shifted in the other direction by the distance -t.

If we take into account all the incident beams contained in the incident cone, **the pairs of transmitted and diffracted points form a transmitted and a diffracted disk.**

The diameter Φ these two disks is directly related to the convergence angle by the relationship:

$$\Phi = 2f\tan\alpha \approx 2f\alpha \qquad \text{(IV.1)}$$

Their centres are separated by the distance:

$$D = f\tan2\theta_B \approx 2f\theta_B \qquad \text{(IV.2)}$$

This is the same as the distance separating the transmitted and diffracted spots in the case of a parallel incident beam (figure III.2). The fact that we are using a convergent incident beam instead of a parallel beam leads to the replacement of spots by disks.

Under two-beam conditions, the convergent-beam electron diffraction pattern consists of a transmitted disk and an hkl diffracted disk.

This convergent beam pattern can also be regarded as the collection of **spot patterns** that would result from parallel incident beams having all possible orientations within the convergent incident beam. Each spot pattern is shifted with respect to the centres of the transmitted and diffracted disks by a distance t, which is a function of the tilt angle ρ (figure IV.3).

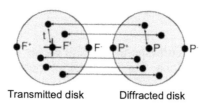

Transmitted disk   Diffracted disk

Figure IV.3 - The convergent-beam diffraction pattern can be considered as made up of a collection of spot patterns.

In the paragraph III.1.2, we indicated that a slight change of the orientation of a parallel incident beam, with respect to the set of (hkl) lattice planes, results in a large modification of the transmitted and diffracted intensities. The convergent beam diffraction pattern can also be regarded as **a map of the transmitted and diffracted intensities according to the orientation of the incident beam with respect to the set of (hkl) lattice planes.**

To describe more precisely the transmitted and the diffracted intensity distributions inside the disks, we can simplify the diagram on figure IV.1. Since all the rays leaving the specimen with the same direction are focused at the same point in the back focal plane of the objective lens, we can reduce or increase the size of the diffracted area without changing the direction of the electron ray-paths. This means that **the diffraction pattern does not depend on the size of the diffracted area** (provided the thickness and the orientation of the lattice planes remain constant in this area).

Taking this property into account, we can replace the diffracted area by a single point E and obtain the simplified diagram shown on figure IV.4. Experimentally, it corresponds to a very small probe. The only disadvantage of this simplified drawing is that it does not show clearly that all the transmitted and diffracted rays are focused on the back focal plane. We can expect that the strongest diffracted intensity will be observed for the incident rays that are exactly at the Bragg orientation for the set of (hkl) lattice planes (in chapter VII, we shall see that this assumption is not always valid).

How do we identify these incident rays?

According to the 3D description of Bragg's law given in chapter I, the incident rays at the Bragg orientation for the (hkl) lattice planes are located on the hkl incident Kossel cone. A part of this Kossel cone, the ABE conical surface, is included inside the convergent electron beam (figure IV.a). Therefore, this surface contains all the incident electrons that are exactly at the Bragg orientation. Since both the convergence 2α and the Bragg angles θ_B are very small in electron diffraction, we can consider this conical surface as planar (figure IV.5b). We note that the $\overline{hkl}$ incident Kossel cone is not involved in this process.

These incident rays at the exact Bragg orientation produce a diffracted line A_DB_D in the back focal plane, which is brighter than the whole diffracted disk. It is called the **excess hkl line**. To produce this excess line, many electrons are removed from the ABE surface. We understand that a deficit of electrons is then produced in the transmitted disk along the line A_TB_T that appears darker than the rest of the disk. This dark line is called the **deficiency hkl line.**

These two lines are parallel to the trace of the set of (hkl) lattice planes. They run through the centre of the transmitted and diffracted disks, and display a separation, which is related to the Bragg angle by the relationship:

$$D \approx 2f\theta_B \tag{IV.3}$$

The appearance of these lines will be described in detail in chapter VII.

Note

In Convergent Beam Electron Diffraction, the transmitted disk is usually called the **Bright-Field** *(BF) disk. The diffracted disk is called the* **Dark-Field** *(DF) disk when the excess hkl line runs exactly through its centre.*

IV.1.1.2 - Ewald sphere construction

The Ewald sphere construction for a convergent incident beam is given on figure IV.7. It is more complex than for a parallel incident beam since an Ewald sphere must be associated with each incident rays contained within the incident beam.

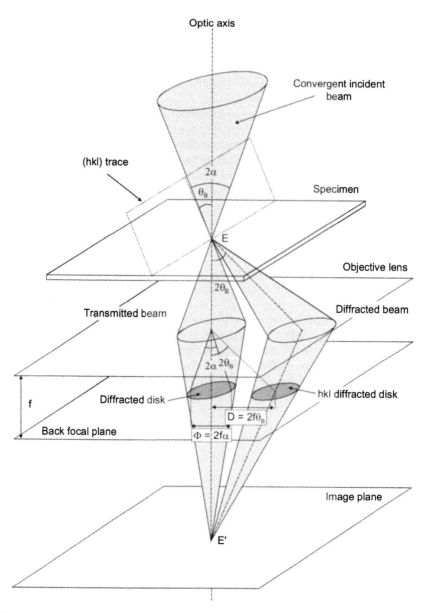

Figure IV.4 - Formation of a convergent-beam electron diffraction pattern. Simplified version of Figure IV.1. The diffracted area is represented by the single point E.

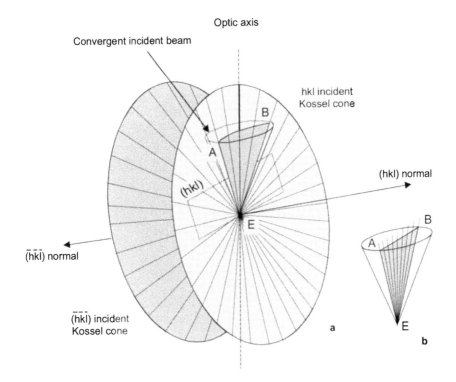

Figure IV.5 - Identification of the incident rays at the exact Bragg orientation.
a - Relative arrangement of the (hkl) incident Kossel cone and of the convergent incident beam. The conical surface ABE contains all the incident electrons at the exact Bragg orientation for the set of (hkl) lattice planes. The $\overline{hkl}$ incident Kossel cone is not involved.
b - The conical surface ABE can be assumed to be a plane taking into account the small value of the Bragg angles θ_B and of the convergent semi-angle α.

For the sake of clarity, the Ewald sphere construction on figure IV.7 is projected on a plane containing the optic axis and the reciprocal vector $\mathbf{g}_{hkl}$. Only three Ewald spheres are drawn corresponding to the three rays "$-\alpha$", "zero" and "$+\alpha$" shown on figure IV.1. These three spheres intersect the hkl relrod between the two limiting points A and B.

The "zero" sphere, related to the incident ray directed along the optic axis, intersects the hkl relrod exactly at the node P. It corresponds to s = 0, meaning that the set of (hkl) lattice planes is at the exact Bragg orientation at point O on figure IV.7 where the diffracted beam OP is produced.

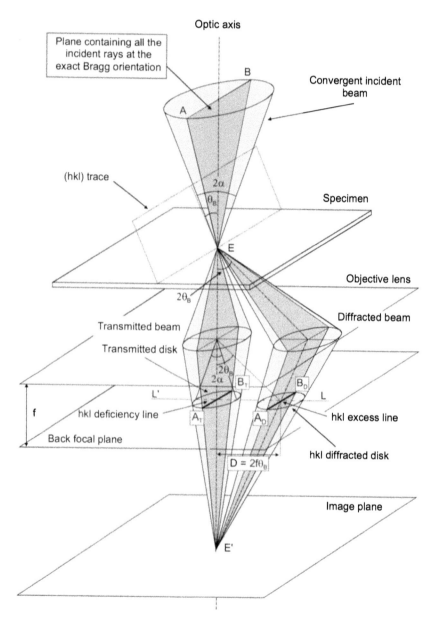

Figure IV.6 - Formation of the convergent beam diffraction pattern. Simplified diagram. The incident plane ABE contains all the incident rays at the exact Bragg orientation. These rays give an hkl excess line ($A_D B_D$ line) located inside the diffracted disk and an hkl deficiency line ($A_T B_T$ line) located inside the transmitted disk.

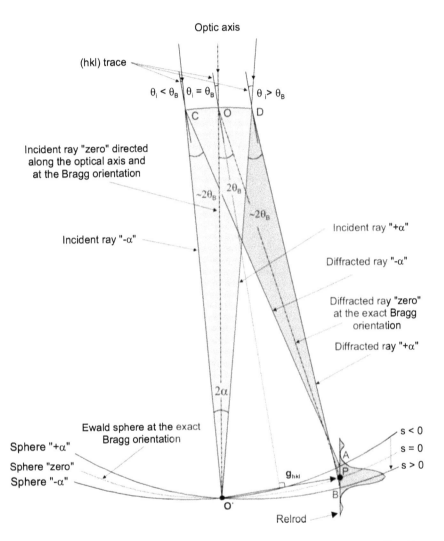

Figure IV.7 - Ewald sphere construction in convergent beam electron diffraction. Exact two-beam conditions.
An Ewald sphere is associated with each incident ray. Only three spheres corresponding to the incident rays "zero" directed along the optic axis, "-α" and "+α" directed along the two opposite extreme directions are drawn. These spheres produce the diffracted beams OP, CA and DB.

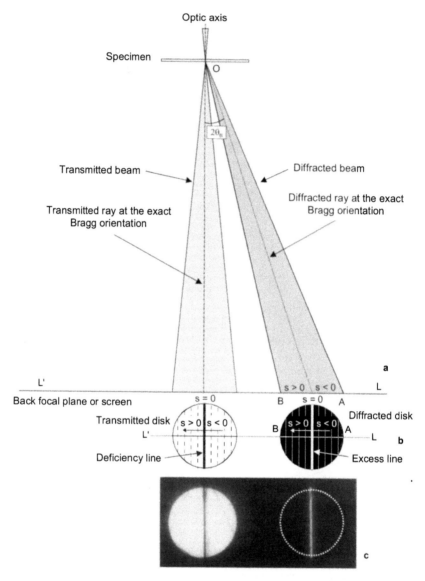

Figure IV.8 - Convergent-beam electron diffraction. Exact two-beam conditions.
a - Relationship between the Ewald sphere construction and the diffraction pattern. All the transmitted and diffracted beams are redrawn starting from a common origin O.
b - Variation of the deviation parameter s inside the diffracted disk. This variation occurs along lines parallel to the s = 0 line, which runs through the centre of the disk. The same variation occurs inside the transmitted disk.
c - Experimental diffraction pattern obtained under exact two-beam conditions. The hkl excess and deficiency lines run through the centre of the diffracted and transmitted disks.

The "-α" sphere, which runs through point A of the relrod, corresponds to a negative deviation parameter (s < 0 since the hkl node is located outside this sphere). At point C, the incident ray "-α" forms an angle $\theta_i < \theta_B$ with respect to the (hkl) lattice planes and gives the diffracted ray CA.

For the "+α" sphere, which runs through point B, the deviation parameter s is positive. θ_i is larger than θ_B at point D and the diffracted beam is directed along D_B.

If we extend this description to the whole set of Ewald spheres located between the two limiting spheres "-α" and "+α", the relrod is cut between the two limiting points A and B.

The Ewald sphere construction gives the directions of the transmitted and diffracted rays corresponding to each incident ray. However, the connection with the diffraction pattern is not as straightforward as in the case of a parallel incident beam (figure III.6) because the transmitted and diffracted rays have different origins (points O, C and D on figure IV.7). To overcome this difficulty, we need to consider figure IV.8, where the transmitted and diffracted rays are redrawn from a common origin O while preserving their orientation.

The two previous figures are two-dimensional representations of the Ewald sphere construction. They only give the transmitted and diffracted intensities along a diameter of the disks parallel to the vector g_{hkl} (along the line LL' drawn on figure IV.6). If we extend this analysis to a three-dimensional construction, we obtain figure IV.8b, which shows that the deviation parameter s varies inside the diffracted disk along parallel lines perpendicular to g_{hkl}. Moreover, the lines with s > 0 and s < 0 are symmetrically disposed with respect to the s = 0 line, which runs exactly through the centre of the diffracted disk. Henceforth, the variation of the deviation parameter inside the diffracted disk will be represented by an arrow directed towards the positive direction of s. The same variation of s occurs in the transmitted disk.

The maximum of diffracted intensity and the minimum of transmitted intensity are observed for the value s = 0 giving the excess and deficiency hkl lines mentioned previously. An example of an experimental pattern obtained under these conditions is given on figure IV.8c.

From the point of view of the Ewald sphere construction, the convergent-beam diffraction pattern is regarded as a map of the **transmitted and diffracted intensities versus the deviation parameter s**.

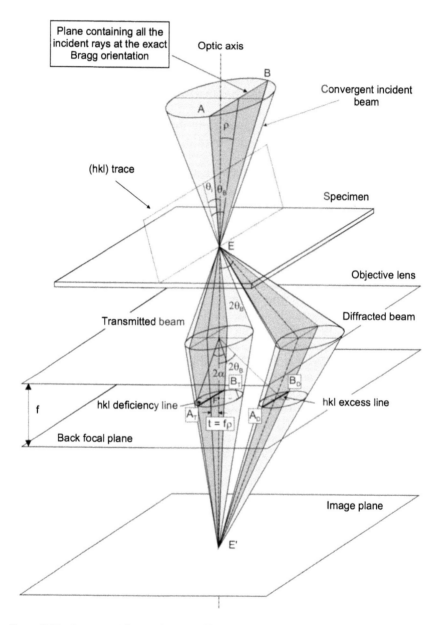

Figure IV.9 - Convergent-beam electron diffraction. Near two-beam conditions. The incident rays located in the plane ABE are at the exact Bragg orientation for the set of (hkl) lattice planes. These rays give an hkl excess line and an hkl deficiency line; these do not run through the centre of the diffracted and transmitted disks.

It is also an intensity map of the transmitted and diffracted intensities as a function of the orientation of the incident beam with respect to the set of (hkl) lattice planes.

In chapter VII in which the intensity of the diffracted beams is studied, we shall see that the line L'L is a direct representation of the "rocking curves", which can be calculated using the kinematical and dynamical theories.

IV.1.2 - A set of (hkl) lattice planes is near the Bragg orientation: near two-beam conditions

IV.1.2.1 - Formation of the diffraction pattern

If the set of (hkl) lattice planes is not exactly at the Bragg orientation for the incident ray directed along the optic axis, it may nevertheless be at the exact Bragg position for incident rays located in an incident plane that does not contain this axis. This is the case for the rays located in the incident plane ABE (which is not vertical) on figure IV.9. As said previously, these incident beams produce an hkl excess line $A_D B_D$ and an hkl deficiency line $A_T B_T$. These lines are separated by the distance $D = 2f\theta_B$, but they do not run through the centre of the corresponding disks. Their position in the disks depends on the orientation of the set of (hkl) lattice planes with respect to the optic axis. Actually, the deficiency line is shifted from the back focal point F of the objective lens by a distance $t = f\rho = f(\theta_I - \theta_B) = f \Delta\theta$.   (IV.4)

IV.1.2.2 - Ewald sphere construction

The corresponding Ewald sphere construction (figure IV.10a) shows that the two points of intersection A and B of the hkl relrod with the limiting spheres "$+\alpha$" and "$-\alpha$" are no longer symmetric about the hkl node. The value $s = 0$ is not at the centre of the diffracted and transmitted disks (figure IV.10b). The experimental pattern of figure IV.10c illustrates this property.

IV.1.3 - Particular case: the set of (hkl) lattice planes is parallel to the optic axis

This particular case deserves to be examined in detail because it occurs in multi-beam zone-axis patterns, which will be described in chapter IV.2.2.

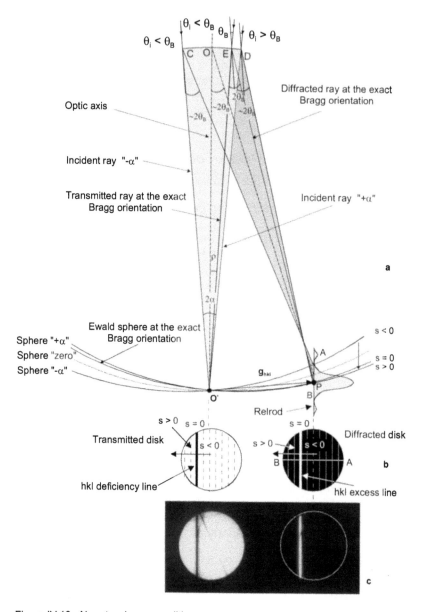

Figure IV.10 - Near two-beam conditions.
a - Ewald sphere construction.
b - Variation of the deviation parameter s inside the diffracted and transmitted disks.
c - Example of an experimental diffraction pattern obtained under the near two-beam conditions. The excess and deficiency lines do not run through the centre of the disks.

When a set of (hkl) lattice planes is parallel to the optic axis, we observe the same result as in the case of a parallel incident beam (chapter III.1.3.2), except for the spot reflections, which are here replaced by disks.

- The reflections g_{hkl} and $g_{\bar{h}\bar{k}\bar{l}}$ are simultaneously excited (three-beam conditions),

- The line s = 0 is not located inside the diffracted disk but halfway between the transmitted and the diffracted disks (figure IV.11).

(220) trace

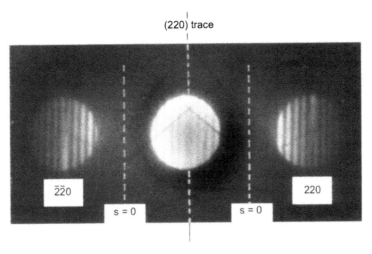

Figure IV.11 - Experimental diffraction pattern from a silicon specimen. The set of (220) lattice planes is parallel to the optical axis and to the incident cone. The value s = 0 is situated halfway between the transmitted and diffracted disks. Fringes are observed inside the disks because the diffracted beams have a very small extinction distance (ξ_g = 103 nm at 300 kV). The sharp lines visible inside the disks come from other lattice planes.

Under these conditions, no excess and deficiency line should be observed inside these disks. This is indeed the case with weak reflections. On the other hand, strong reflections often produce fringes. They have a large structure factor F_{hkl} and thus a short extinction distance ξ_g. We shall see, in chapter VII, that the corresponding nodes of the reciprocal lattice are then large relrods, which display strong intensity modulation and produce fringes. This means that diffracted intensity can be observed over a very broad angular field on both sides of s = 0 and fringes then appear inside the disks. The experimental pattern of figure IV.11 is an example of this. Fringes are clearly visible in the 220 and $\overline{2}\overline{2}0$ silicon disks for which the extinction distance ξ_g is 103 nm at 300 kV.

IV.1.4 - The incident beam is tilted with respect to the optic axis

As in the case of spot patterns (see chapter III.1.3.1), the misorientation of the convergent incident beam by an angle ρ with respect to the optic axis results in a shift of both the transmitted and diffracted disks by a distance $t = f\rho$ (figure IV.12). Note that, during this operation, the excess and deficiency lines remain motionless because their position depends only on the orientation of the lattice planes with respect to the optic axis.

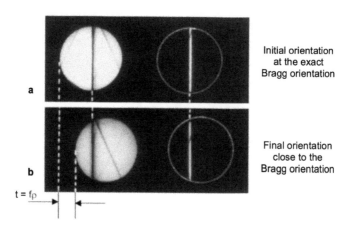

Figure IV.12 - Effect of a beam misorientation ρ on the diffraction pattern.
a - Initial position. The exact Bragg orientation is obtained when the incident beam is parallel to the optic axis.
b - The incident beam is tilted by an angle ρ with respect to the optic axis. The disks are shifted by a distance $t = f\rho$. The positions of the excess and deficiency lines remain unchanged because they are "attached" to the specimen, which does not move. The lines no longer run through the centres of the disks.
The few sharp lines visible inside the transmitted disks are deficiency lines connected with other lattice planes.

IV.1.5 - Application: setting (hkl) lattice planes at the exact Bragg orientation

As indicated above, the misorientation of a set of (hkl) lattice planes with respect to the convergent incident beam results in a shift of the excess and deficiency hkl lines inside the transmitted and diffracted disks. These lines are "attached" to the specimen and move with it.

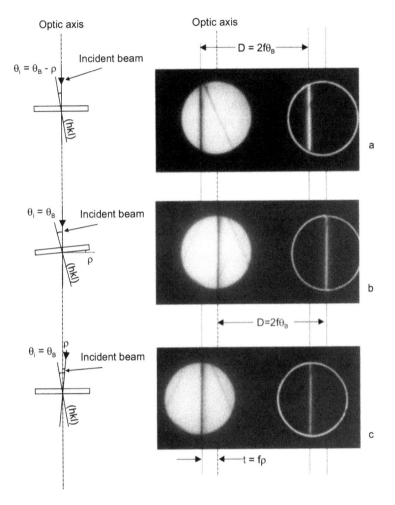

Figure IV.13 - Setting (hkl) lattice planes at the exact Bragg position starting from an initial approximate orientation.
a - Initial approximate orientation. The excess and deficiency lines do not run through the centre of the transmitted and diffracted disks.
b - The specimen is tilted by an angle ρ until the incidence angle $θ_i$ equals the Bragg angle $θ_B$. The lines are shifted by a distance t = fρ and run through the centre of the disks. Note that the position of the disk remains unchanged.
c - The incident beam is tilted with respect to the optical axis by an angle ρ until the incidence angle $θ_i$ equals the Bragg angle $θ_B$. The disks are shifted by the distance t = fρ. The position of the lines remains unchanged, and they run through the centre of the disks.

Moreover, this displacement is very sensitive and can be used to position accurately the crystal in a given orientation. This interesting property has several applications.

For example, we can use it to position a diffracted disk at the exact Bragg orientation, setting out from an initial position for which the Bragg orientation is only approximate (figure IV.13a). The method consists in tilting the specimen through an angle ρ until the excess and deficiency lines run exactly through the centre of the transmitted and diffracted disks (figures IV.13a and b).

Instead of tilting the specimen, we can also tilt the incident beam by an angle ρ using the microscope beam deflection coils (the ones used to obtain conventional dark field images) until $\theta_I = \theta_B$. This alternative method, illustrated on figure IV.13c, does not produce exactly the same effects at the diffraction pattern. In this second case, the hkl excess and deficiency lines remain motionless, because they are "attached" to the specimen. The transmitted and diffracted disks are now shifted with the incident beam to a distance $t = f\rho$. The disks are now "attached" to the incident beam.

Note that this effect is not observed with spot patterns. Indeed, we indicated in chapter III.1.4 that a slight change of the orientation of the (hkl) lattice planes does not produce any significant modification of the geometry of the spot pattern. Only the intensity of the reflections is strongly affected.

IV.2 - Diffraction pattern under many-beam conditions

IV.2.1 - General case

Usually, several sets of (hkl) lattice planes can simultaneously satisfy the Bragg orientation, exactly or approximately, and the diffraction pattern then displays simultaneously several diffracted disks (figure IV.14). **Each hkl diffracted disk contains its hkl excess line.** This line runs through the centre of the diffracted disk if the Bragg orientation is exactly obtained. It is almost the case for the line located in the disk A of the experimental pattern shown on figure IV.14a.

The transmitted disk contains all the corresponding deficiency lines.

We note that, for each pair of hkl excess and deficiency lines,
- the lines are parallel,
- they are separated by a distance $D = 2f\theta_B$,

- the trace of the (hkl) lattice planes is located halfway between the two lines (figure IV.1a),
- they are located at equivalent positions in the diffracted and in the transmitted disk.

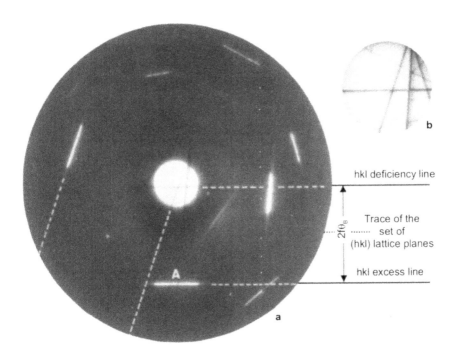

Figure IV.14 - Diffraction pattern under many-beam conditions.
a - Whole pattern. Each set of (hkl) lattice planes produces an excess line situated inside its diffracted disk and a deficiency line located inside the transmitted disk. These two lines are separated by a distance D = 2fθ and the lattice plane trace is situated halfway between the transmitted and diffracted disks. The transmitted disk contains all the deficiency lines.
The weak lines located outside the disks in the prolongation of the excess and deficiency lines are Kikuchi lines.
b - Enlargement of the transmitted disk.

Note
On the experimental pattern on figure IV.1a, we note the presence of weak lines in the prolongation of the excess and deficiency lines. They are Kikuchi lines, which will be described in chapter VI.

IV.2.1.1 - Influence of the convergence semi-angle α on the number of diffracted reflections

Using a convergent incident beam results in a larger number of reflections in the diffraction pattern than in a spot pattern. This property can be explained by using the Ewald sphere construction shown on figure IV.15.

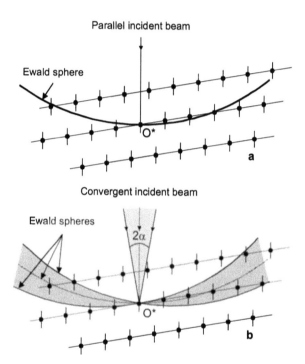

Figure IV.15 - Ewald sphere construction showing the influence of the convergence angle 2α on the number of reflections.
a - Parallel incident beam. A few relrods intersect the single Ewald sphere and produce diffracted beams.
b - Convergent incident beam. Numerous relrods intersect the set of Ewald spheres corresponding to the different incident beams contained in the incident convergent beam.

This shows that the probability for a relrod to intersect the Ewald sphere is greater for the set of Ewald spheres related to a convergent incident beam (figure IV.15b) than for the single Ewald sphere related to a parallel incident beam (figure IV.15a). We shall see that the very large beam convergences used to form the Kossel or the LACBED patterns will produce a large number of diffracted beams.

IV.2.2 - Particular case: [uvw] zone-axis pattern

This particular case is very significant because of its numerous applications and it deserves a detailed study.

The description in terms of **Laue zones** given for a parallel incident beam (chapter III.2.2) also applies for a convergent incident beam. The spot reflections are simply transformed into disks, which contain the excess and deficiency lines (figures IV.16).

The disks located in the zero-order Laue zone correspond to g_{hkl} vectors of the zero layer of the reciprocal lattice. They are thus perpendicular to the incident beam and correspond to lattice planes parallel to this beam (these planes are vertical in the microscope). For these vertical planes, the value $s = 0$ for the excess hkl line is not located inside the disk but halfway between the centre of the transmitted and diffracted disks (see paragraph IV.1.3). **No excess or deficiency line should be observed in the diffracted and transmitted disks for the Zero-Order Laue Zone.** Actually, more or less complex fringe systems are often observed inside these disks.

The origin of these fringes can be explained in the following way:
- the reflections belonging to the Zero-Order Laue Zone are usually strong and have small extinction distance ξ_g. Under three-beam conditions (see paragraph IV.1.3) these reflections display a modulation of the diffracted intensity within a very broad angular field, which results in the presence of a system of bright and dark fringes. In the multi-beam conditions, each individual reflection of the Zero-Order Laue Zone is under three-beam conditions and produces its own system of fringes. These systems interact to create the complex patterns observed. In the transmitted disk, we observe typical figures made up of concentric fringes whose symmetry is related to the nature of the zone axis (Figure IV.16b). The number of fringes depends mainly on the thickness of the illuminated zone and can be used to measure the specimen thickness. On the other hand, the symmetry of the fringe system does not depend on the thickness.

We note that these broad fringes give only two-dimensional information since they originate from the zero layer of the reciprocal lattice. For this reason, their symmetry is described as "projection" symmetry.

We shall describe these features in detail in chapter VII, where dynamical interactions are dealt with.

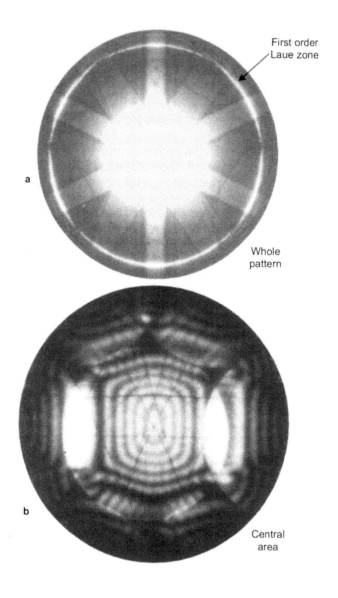

Figure IV.16 - [114] zone-axis pattern from a silicon specimen.
a - Whole pattern. The excess lines of the first-order Laue zone are located on a circle.
b - Enlargement of the central area of the pattern. The transmitted disk contains sharp HOLZ lines as well as a system of broad and concentric fringes.

The disks located in the High-Order Laue Zones correspond to reciprocal layers with $n \geq 1$. Each diffracted hkl disk contains its excess hkl line. This is clearly visible in the experimental pattern on figure IV.16a. We shall explain, in chapter VII, that these lines are usually very sharp and weak because they correspond to reflections having a large extinction distance ξ_g.

For each Laue zone, the excess lines are all located on a circle of radius R_n, which corresponds to the intersection of the n-th order reciprocal layer with the Ewald sphere. The radius R_n of this circle is related to the lattice parameter $P_{[uvw]}$ of the [uvw] zone axis by:

$$R_n = (2nH/\lambda - n^2H^2)^{1/2} \tag{IV.5}$$

with $H = 1/P_{[uvw]}$ (III.20)

For a zone-axis diffraction pattern, the excess lines are called **HOLZ lines** (High-Order Laue Zone).

Note

In practice, the fine structure of the excess lines in the High-Order Laue Zones can be more complex than the simple description given here. This is the case for the diffraction pattern of figure IV.17, which shows a splitting of the excess lines into two or three lines. This effect depends on the zone axis and on the nature of the specimen. It is due to the multiple dynamical interactions and can be explained by means of dispersion surfaces.

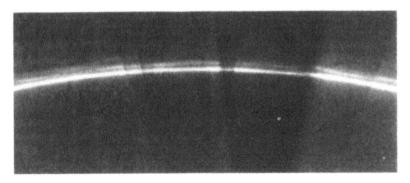

Figure IV.17 - Fine structure of the excess lines observed in the first-order Laue zone of a silicon specimen. The lines are split. This effect depends on the zone axis and on the nature of the specimen. It is related to multiple dynamical interactions.

We have already indicated that the transmitted disk contains broad fringes caused by interactions occurring in the Zero-order Laue zone. **It also contains all the deficiency lines connected with the excess lines located in the High-Order Laue Zones.** They are also called **HOLZ lines.**

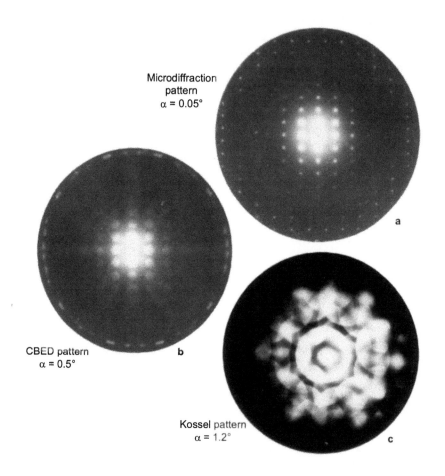

Microdiffraction
pattern
$\alpha = 0.05°$

a

CBED pattern   b
$\alpha = 0.5°$

Kossel pattern
$\alpha = 1.2°$ c

Figure IV.18 - Influence of the convergence semi-angle α on the diffraction pattern. <310> zone-axis pattern from a ferrite specimen.
a - Microdiffraction pattern. The convergence semi-angle α is about 0.05°. The disks have a very small diameter.
b - CBED pattern. The convergence semi-angle α is about 0.5°. The disks do not overlap.
c - Kossel pattern. The convergence semi-angle α is about 1.2°. The disks overlap strongly.

These lines are usually sharp and very sensitive to lattice parameter variations and to voltage variations. On the other hand, they do not depend significantly on the specimen thickness. They are very useful for numerous applications: crystal point-group identification, accurate lattice parameter measurement, local deformation measurement, crystal orientation identification...

HOLZ lines are clearly seen on the experimental pattern on figure IV.16b. Note that these lines give three-dimensional information since they involve several reciprocal layers.

IV.3 - Influence of the convergence semi-angle α on the CBED pattern

The diameter of the transmitted and diffracted disks is related to the convergence semi-angle α of the incident beam. Depending on the disk diameter, three types of diffraction patterns can be distinguished:

- when the disks have a very small diameter (the incident beam is almost parallel), the diffraction pattern looks like a conventional selected-area electron diffraction pattern and is called a **microdiffraction pattern** (figure IV.1a).

- when the diffracted disks are large but do not overlap (figure IV.18b), true **convergent-beam patterns** are obtained. The convergence semi-angle α is about 0.5°.

- when the disks have a very large diameter and overlap to some extent (figure IV.18c), the patterns are called **Kossel patterns**. The beam convergence semi-angle α can reach 3° to 5° with modern microscopes. Unfortunately, Kossel patterns display a poor quality in the superimposed areas. We explain why in the following chapter.

CHAPTER V
Diffraction pattern produced by a large-angle convergent incident beam: Kossel pattern

V.1 - Advantage of a large convergence angle

In the previous chapter, we indicated that the bright-field disk (the transmitted disk) of a CBED pattern contains deficiency lines. The larger the convergence angle, the larger the diameter of the transmitted disk and the larger the number of lines visible inside this disk. This effect is illustrated on figure V.1. Since a large number of lines allows easier and more accurate interpretations and measurements, it is sensible to use the largest available convergence angle. This angle is about two to five degrees with modern microscopes.

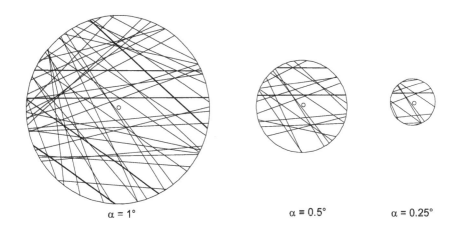

$$\alpha = 1° \qquad\qquad \alpha = 0.5° \qquad \alpha = 0.25°$$

Figure V.1 - Simulations of the transmitted disk of a convergent beam diffraction pattern for different values of the convergence semi-angle α. The larger the convergence angle, the larger the number of deficiency lines observed.

V.2 - Kossel patterns

V.2.1 - General case

Consider the simplified diagram of figure V.2. It displays the same experimental conditions as the ones on figure IV.4 except that the convergence angle of the incident beam is now a few degrees. For the sake of clarity, this angle is very strongly exaggerated.

The transmitted and the (hkl) diffracted rays give two overlapping disks in the back focal plane of the objective lens. These disks overlap when **the convergence semi-angle α is larger than the Bragg angle θ_B** (since the distance between the centre of the two disks is $D = 2f\theta_B$ and their diameter is $\Phi = 2f\alpha$).

The convergence of the incident beam is now much larger than the Bragg angles. As a result, this beam can also contain incident rays at the Bragg orientation on the other side of the set of lattice planes, i.e. for the $(\bar{h}\bar{k}\bar{l})$ lattice planes. These rays give a second $\bar{h}\bar{k}\bar{l}$ diffracted disk, which also overlaps the transmitted disk.

We have to take into account the (hkl) and $(\bar{h}\bar{k}\bar{l})$ incident Kossel cones (see chapter I) in order to predict the intensity distribution inside these three disks (Figure V.3). The conical surfaces ABE and CDE of these Kossel cones are included in the incident convergent beam. They contain all incident electrons that are exactly at the Bragg orientation for the (hkl) and $(\bar{h}\bar{k}\bar{l})$ lattice planes and can be regarded as planar surfaces since the Bragg and convergence angles are so small.

According to the description given in chapter IV.1.1.1, all the incident rays on the ABE surface give a bright hkl excess line ($A_D B_D$ line) located in the hkl diffracted disk, and a dark hkl deficiency line ($A_T B_T$ line) located in the transmitted disk (Figure V.4).

As in the case of the CBED patterns:
- the centres O_T and O_{D+} of the disks are separated by $D = 2f\theta_B$,
- the same distance D also separates the excess and deficiency hkl lines,
- the lines are separated from the centre of their respective disk by the distance $t = f\rho$.

In the same way, the incident rays located in the CDE plane produce an $\bar{h}\bar{k}\bar{l}$ excess line ($C_D D_D$ line) located in the $\bar{h}\bar{k}\bar{l}$ diffracted disk and an $\bar{h}\bar{k}\bar{l}$ deficiency line ($C_T D_T$ line) located in the transmitted disk (Figure V.5).

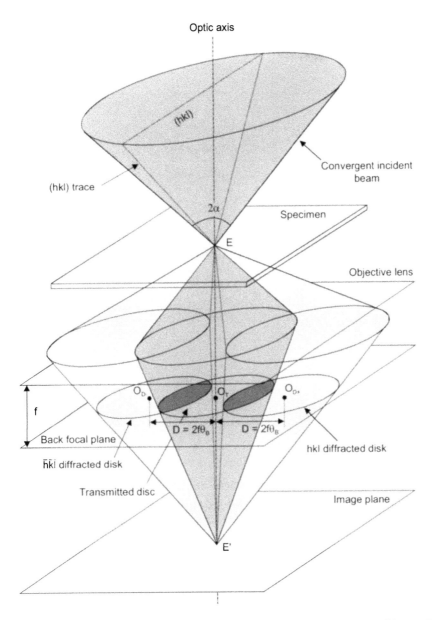

Figure V.2 - Kossel pattern. Electron ray-paths of the diffracted and transmitted beams for a set of (hkl) lattice planes. Owing to the large convergence of the incident beam, the hkl and $\overline{hkl}$ diffracted disks are partially superimposed on the transmitted disk in the back focal plane of the objective lens.

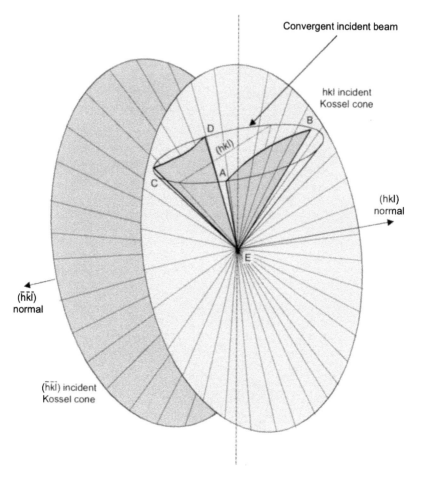

Figure V.3 - Identification of the incident rays at the exact Bragg orientation.

a - Relative positions of the (hkl) and ($\overline{h}\overline{k}\overline{l}$) incident Kossel cones with respect to the incident convergent electron beam. The conical surfaces ABE and CDE contain all incident electrons that are at the exact Bragg orientation for the (hkl) and ($\overline{h}\overline{k}\overline{l}$) lattice planes respectively. Taking into account the low values of the convergence and Bragg angles α and θ_B, the two conical surfaces ABE and CDE can be regarded as planes.

The hkl transmitted Kossel cone and the $\overline{h}\overline{k}\overline{l}$ diffracted Kossel cone are identical (see chapter I), as are the $\overline{h}\overline{k}\overline{l}$ transmitted and the hkl diffracted Kossel cones (Figure I.6). This means that the hkl excess line (A_DB_D line) and the $\overline{h}\overline{k}\overline{l}$ deficiency line(C_TD_T line) are superimposed in the area of overlap of the hkl diffracted and transmitted disks.

The same situation occurs for the $\overline{hkl}$ excess line ($C_D D_D$ line) and the hkl deficiency line ($A_T B_T$ line). This effect is clearly visible on figure V.6, which represents a cross-section through figures V.4 and V.5 along a vertical plane containing the point E.

In both cases, one line displays an excess of electrons while the other displays a deficiency of electrons. The overall contrast tends to cancel out and to produce a poor quality pattern as shown on figure V.7.

Note
On the experimental pattern of figure V.8, the contrast is very poor in the central area of the transmitted disk where the superimposition of the excess and deficiency lines is nearly perfect. The contrast is better at the periphery of the pattern because the superimposition of the lines is not perfect. In chapter VIII, we show that this poor superimposition is related to the very strong spherical aberration of the objective lens.

This analysis involves a single set of (hkl) lattice planes. Because of the large convergence angle of the incident beam, many sets of lattice planes can be simultaneously at the Bragg orientation and several diffracted beams are obtained. Each of them gives an excess line located in its corresponding diffracted disk and a deficiency line located in the transmitted disk (Figure V.7). Thus, a Kossel pattern always contains a very large number of lines. This number increases with the convergence angle of the incident beam (see paragraph IV.2.1.1).

V.2.2 - [uvw] zone-axis Kossel pattern

When a [uvw] zone axis of the crystal is parallel to the incident cone, the Kossel pattern (Figure V.8) contains several lines disposed radially around the centre of the pattern. These lines correspond to (hkl) lattice planes in the zone with the [uvw] zone axis.

V.2.3 - Main characteristics of the Kossel patterns

The lines can be grouped into pairs of parallel lines.
- If the two lines of a pair are both located inside the transmitted disk (Figure V.9a), their contrast is very poor owing to the superimposition mentioned previously. This is the case for the lines I and I' on figure V.7.
- If one line of a pair is located inside the transmitted disk (a dark deficiency line) and the other one inside a diffracted disk (a bright excess line) (Figure V.9b), their contrast is usually good since there is no superimposition. This is the case for the lines II and II', III and III' or IV and IV' on figure V.7.

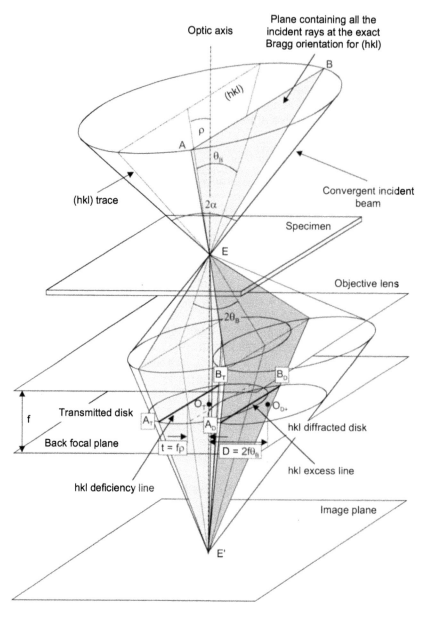

Figure V.4 - Kossel pattern. Electron ray-path of the diffracted and transmitted rays for a set of (hkl) lattice planes. The incident electrons located in the ABE plane are at the exact Bragg orientation for the set of (hkl) lattice planes. They produce the excess and deficiency hkl lines located in the back focal plane of the objective lens.

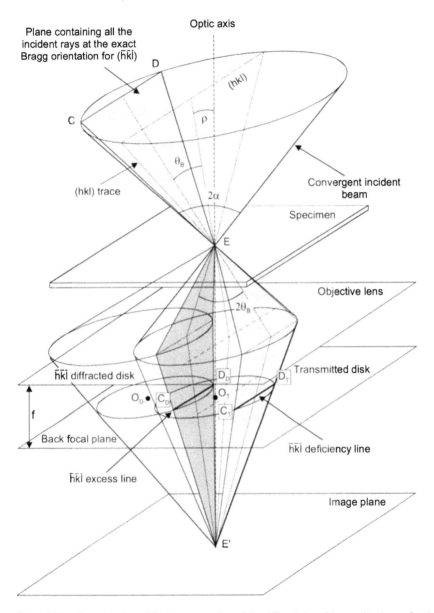

Figure V.5 - Kossel pattern. Electron ray-paths of the diffracted and transmitted rays for the set of $(\bar{h}\bar{k}\bar{l})$ lattice planes. The incident electrons located in the CDE surface are at the exact Bragg orientation for the set of $(\bar{h}\bar{k}\bar{l})$ lattice planes. They produce the excess and deficiency $\bar{h}\bar{k}\bar{l}$ lines located in the back focal plane of the objective lens.

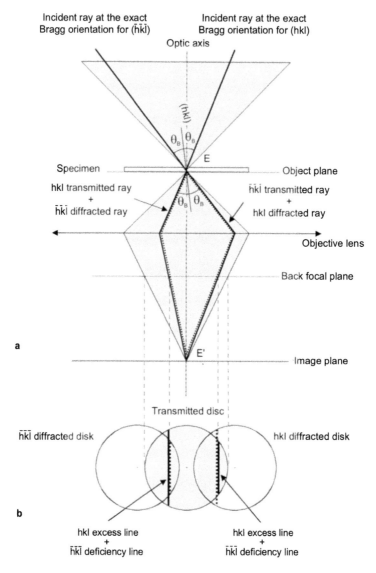

Figure V.6 - Kossel pattern. Vertical cross-section through figures V.4 and V.5.

a - Electron ray-paths at the exact Bragg orientation for the sets of (hkl) and ($\bar{h}\bar{k}\bar{l}$) lattice planes.

b - Detail of the superimposition of the excess and deficiency lines in the overlapping areas of the hkl and $\bar{h}\bar{k}\bar{l}$ diffracted disks with the transmitted disk.

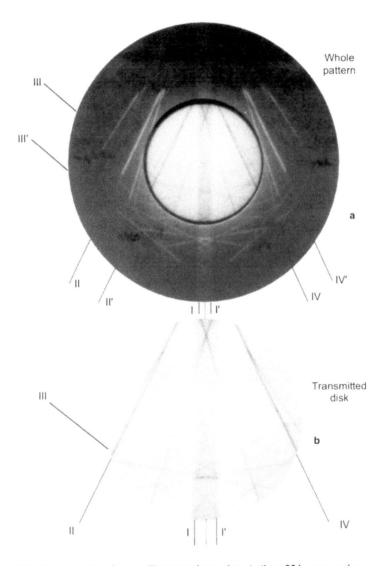

Figure V.7 - Kossel pattern from a silicon specimen close to the <001> zone axis.
a - Whole pattern. A montage was carried out in order to compensate for the strong brightness differences between the central and the peripheral areas of the pattern.
b - Enlargement of the transmitted disk.

Figure V.8 - Transmitted disk of an experimental <001> zone-axis Kossel pattern from a silicon specimen. Owing to the strong spherical aberration of the magnetic lenses, the superimposition of the deficiency and excess lines is imperfect.

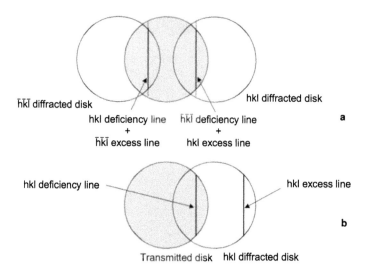

Figure V.9 - Schematic description of the two possible locations of a pair of lines on Kossel patterns.
a - The two lines of the pair are located inside the transmitted disk. They are superimposed and their contrast is poor.
b - One line of the pair (the hkl deficiency line) is located inside the transmitted disk and the other (the excess hkl line) is located inside the hkl diffracted disk. The two lines are not superimposed and their contrast is good.

For each pair of lines (Figure V.10):
- the separation between the lines is directly related to the Bragg angle by the relationship:

$$D = 2f\theta_B \qquad \qquad \text{(V.1)}$$

- the trace of the set of (hkl) lattice planes is located halfway between the two lines,
- the trace of a set of lattice planes running through the centre of the transmitted disk means that the planes are parallel to the incident beam axis. They are vertical in the microscope (if the incident beam axis is parallel to the optical axis).
- the trace of lattice planes located at a distance $t = f\rho$ from the centre of the transmitted disk corresponds to lattice planes tilted at an angle ρ with respect to the optic axis.
- two or several traces can intersect at a point, which corresponds to a [uvw] zone axis.

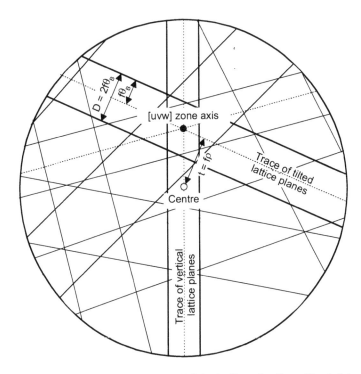

Figure V.10 - Main features of the transmitted disk of a Kossel pattern. The deficiency lines can be sorted into pairs of parallel lines. The traces of the sets of lattice planes are located halfway between each pair of lines. Two or several traces intersect along a [uvw] zone axis.

Note that the deficiency and excess lines belonging to a pair are always located at equivalent positions in the transmitted and diffracted disks (Figures V.9a and b).

V.2.3.1 - Particular case of a [uvw] zone axis Kossel pattern

Many traces run through the centre of the transmitted disk on zone-axis patterns. They correspond to vertical planes in the zone with the [uvw] axis (Figure V.11). The angles Φ between the traces are directly related to the angles between the set of lattice planes. These lines are the equivalent of the reflections present in the Zero-Order Laue Zone on CBED patterns.

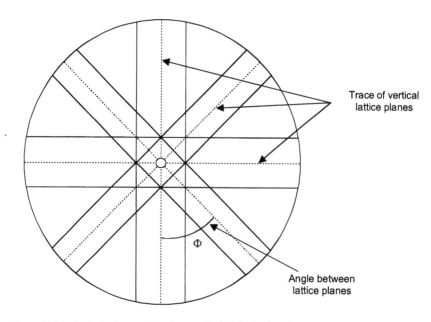

Trace of vertical lattice planes

Φ

Angle between lattice planes

Figure V.11 - Main features of the transmitted disk of a [uvw] zone-axis Kossel pattern. The traces of the sets of lattice planes run through the centre of the pattern and correspond to lattice planes in the zone with the [uvw] zone axis. The angles Φ between the traces correspond to the angles between lattice planes.

V.2.4 - Detail of the superimposition of the hkl deficiency and $\overline{hkl}$ excess lines

We have explained the origin of the superimposition of the excess and deficiency lines observed in the transmitted disk of the Kossel patterns. However, all the lines of a Kossel pattern are not always superimposed. Only a few lines are superimposed on the experimental pattern shown on figure V.7. These lines are not easy to detect because they have a very weak contrast. Most of the lines are not superimposed and display a better contrast. On the other hand, almost all the lines are superimposed on the zone axis pattern of figure V.8. The contrast of the whole pattern is very poor, especially in the central area where the superimposition of the lines is nearly perfect.

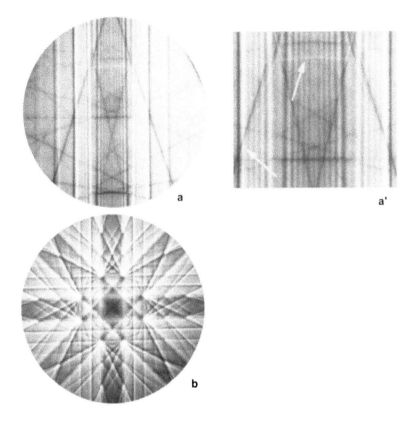

Figure V.12 - Kossel patterns from a silicon specimen obtained by adjusting the diffraction lens slightly to observe a plane above the back focal plane of the objective lens. The deficiency and excess lines are not perfectly superimposed.
a and a' - Transmitted disk (bright field) and enlargement for an orientation close to the <001> zone axis.
b - <001> zone-axis pattern.
Compare these two patterns with the ones shown on figures V.7b and V.8.

It is possible to identify the superimposed lines more easily by adjusting the diffraction lens, thereby observing a plane located slightly above or below the back focal plane of the objective lens where the superimposition is imperfect. The experimental patterns of figure V.12 illustrate this property.

When are the lines superimposed?
Two lines are superimposed when both are located inside the transmitted disk (figure V.13).

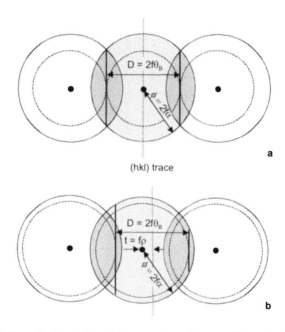

Figure V.13 - Schematic description of the superimposition conditions for the lines present in the transmitted disk of a Kossel pattern.
a - The trace of the set runs through the centre of the transmitted disk. The superimposition of the lines occurs as soon as the disks are tangential (these disks are drawn in dotted lines). The superposition is located in the middle of the superimposed area.
b - The trace of the set of lattice planes does not run through the centre of the transmitted disk. The superposition is not located in the middle of the superimposed area.

Two cases are possible according to whether or not the trace of the set of lattice planes runs through the centre of the transmitted disk.
When the trace runs through the centre of the disk, the superimposition of the lines occurs as soon as:
$$\alpha > \theta_B \hspace{4cm} (V.2)$$

i.e., when the disks are tangential since the distance between the centres of the transmitted and diffracted disks is $D = 2f\theta_B$. The superimposition of the lines occurs in the middle of the overlapping area (Figure 13a).

On a zone-axis pattern, all the planes in the zone have their trace running through the centre of the transmitted disk. Since most of these planes usually have small Bragg angles, they satisfy the previous superimposition condition. This explains why most of the lines are superimposed on figure V.8.

For a set of lattice planes tilted at an angle ρ with respect to the axis of the incident convergent beam, the trace of the set does not run through the centre of the disk. The superimposition condition is:

$$\alpha > \rho + \theta_B \tag{V.3}$$

and the superimposition is no longer located at the centre of the overlapping zone (Figure V.13b).

V.3 - Limiting value of the convergence semi-angle α_{lim}

The superimpositions of the lines occur as soon as the disks overlap, *i.e.* when the convergence semi-angle α becomes larger than a **limiting value** α_{lim}. This value corresponds to the smallest Bragg angle θ_{Bmin} present on the diffraction pattern (Figure V.14).

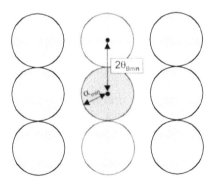

Figure V.14 - Identification of the limiting value α_{lim} of the convergence semi-angle. It occurs when the disks are tangential, i.e. when $\alpha_{lim} = \theta_{Bmin}$.

Bragg's law $\lambda = 2d_{hkl}\sin\theta_B$ states that the limiting angle α_{lim} is a function of the lattice parameters. Crystals with large lattice parameters have large interplanar spacings and produce small Bragg angles. They only allow small limiting values for the convergence angle. This effect is clearly shown on the [001] simulations for silicon with lattice parameter a = 0.543 nm and for the tetragonal σ phase with lattice parameter a = 0.879 nm and c = 0.4554 nm (figures V.15b and c).

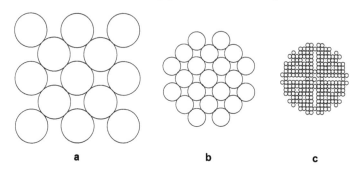

a   **b**   **c**

Figure V.15 - Effect of the lattice parameters and accelerating voltage on the limiting convergence value α_{lim}.
a - Simulated [001] CBED pattern at 100 kV for silicon with lattice parameter a = 0.543 nm. The beam semi-convergence is 0.55°.
b - Simulated [001] CBED pattern at 300 kV for silicon with lattice parameter a = 0.543 nm. The beam semi-convergence is 0.3°.
c - Simulated [001] CBED pattern at 300 kV for σ phase with lattice parameters a = 0.879 nm and c = 0.454 nm. The beam semi-convergence is 0.06°.

How can the limiting value α_{lim} be increased?
We can:

- decrease the accelerating voltage of the electron microscope; the wavelength and, consequently, the Bragg angles are increased. This effect is illustrated on figures V.15a and b, which correspond to a silicon crystal observed at 100 and 300 kV. This experimental method is not very practical and has a limited range.

- find a way of isolating and selecting the transmitted beam from among the diffracted ones. Several solutions have been proposed and will be described in the chapter VIII. The most significant one is the LACBED method developed by Tanaka et al. (i.6), which consists in modifying the height of the specimen in the microscope.

In spite of their poor contrast, Kossel patterns have some specific applications. In the following chapters, we shall see that they can be used to interpret highly distorted LACBED patterns.

CHAPTER VI
Diffraction pattern produced by a large-angle convergent beam: LACBED patterns

VI.1 - Formation of bright- and dark-field LACBED patterns

In this method, the illumination conditions are the same as those used for the Kossel method, i.e. the incident beam has a large convergence and is focused on the object plane of the objective lens. The main difference with the Kossel method concerns the specimen, which is now raised (or lowered) by a distance Δh from its normal position in the object plane (Figure VI.1). We shall show that this modification produces a useful separation of the hkl and $\overline{hkl}$ diffraction phenomena.

For the sake of clarity, only the transmitted and diffracted rays at the exact Bragg orientation ($s = 0$) are shown on figure VI.1.

VI.1.1 - Two-beam conditions

We reconsider the case of a single set of (hkl) lattice planes. As for the Kossel patterns, the incident rays contained in the ABE plane are at the exact Bragg orientation for this set. They pass through the specimen along the line $A_E B_E$, where diffraction occurs, and produce the hkl diffracted rays located in the plane $A_E B_E K$ and the hkl transmitted rays located in the plane $A_E B_E E$. In the same way, the incident rays contained in the plane CDE are at the exact Bragg orientation for the set of $\overline{hkl}$ lattice planes. They pass through the specimen along a second line $C_E D_E$, and produce the $\overline{hkl}$ transmitted and diffracted beams located in the planes $C_E D_E E$ and $C_E D_E L$ respectively.

The fact that the specimen is raised by Δh produces a separation of the two hkl and $\overline{hkl}$ diffraction phenomena, which occur along the two lines $A_E B_E$ and $C_E D_E$ of the specimen. We recall that, in the Kossel method, the two phenomena occurred at the same point E.

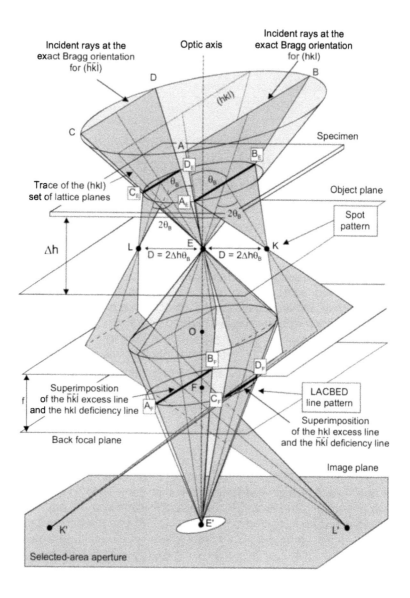

Figure VI.1 - Formation of a LACBED pattern. Setting the specimen at the distance Δh from the object plane of the objective lens produces a spot pattern and a line pattern located in the object plane and in the back focal plane respectively. The transmitted beam is selected by means of the selected-area aperture.

For the sake of clarity, only the transmitted and the diffracted rays at the exact Bragg orientation are shown on this diagram.

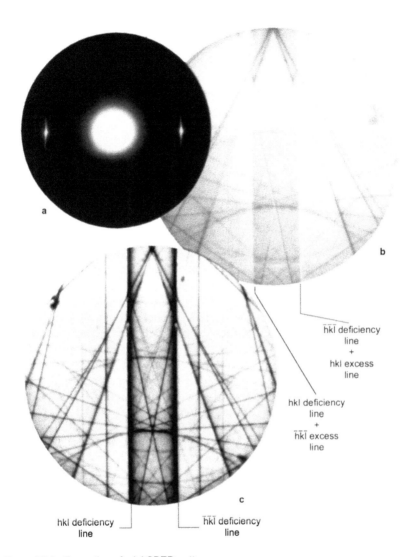

Figure VI.2 - Formation of a LACBED pattern.
a - Spot pattern located in the object and image planes of the objective lens.
b - Line pattern located in the back focal plane of the objective lens. The hkl deficiency line
and the $\overline{hkl}$ excess line are superimposed as are the $\overline{hkl}$ deficiency line and the hkl excess
line. Their contrast is poor.
c - Line pattern located in the back focal plane and observed with a selected-area aperture.
Only the hkl and $\overline{hkl}$ deficiency lines are visible. Their contrast is excellent.

The hkl and $\overline{hkl}$ diffracted beams are separated from the transmitted beam in the object plane, where they produce a spot pattern consisting of the three spots E, K and L. The distance D between the spots E and K and between the spots E and L is related to the Bragg angle θ_B and to the specimen height Δh by:

$$D = 2\Delta h \theta_B \qquad\qquad (VI.1)$$

Figure VI.2a shows an experimental spot pattern. Its reflections are not perfectly symmetrical, and they form trails. This effect results from the strong spherical aberration related to the "nanoprobe" mode used to obtain this pattern. This operating mode will be detailed in chapter VIII, where experimental conditions are dealt with. This spot pattern is magnified by the objective lens in the conjugate image plane (spots K', E' and L' on figure VI.1).

The hkl diffracted rays and the $\overline{hkl}$ transmitted rays intersect the back focal plane of the objective lens along the same line $C_F D_F$. Therefore, this line represents the superimposition of the bright excess hkl line and the dark $\overline{hkl}$ deficiency line. In the same way, the $\overline{hkl}$ diffracted rays and the hkl transmitted rays intersect the back focal plane along a second common line $A_F B_F$. This effect is clearly visible on figure VI.3, which corresponds to a vertical cross-section through figure VI.1. In the back focal plane therefore, there is no difference between a LACBED pattern and a Kossel pattern. This means that **the modification of the specimen height Δh does not produce any change of the line pattern located in the back focal plane of the objective lens**. The excess and deficiency lines remain superimposed as shown on figure VI.2b.

Unlike Kossel patterns, the diffracted and transmitted rays are now separated in the two conjugate object (spots L, E and K) and image (spots L', E' and K') planes of the objective lens. They can be selected and isolated using the selected area-aperture located in the image plane.

VI.1.1.1 - Bright-field pattern

If the transmitted beam is selected, only the two dark deficiency hkl and $\overline{hkl}$ lines are observed (figures VI.1 and VI.3). They are parallel and separated by a distance $D = 2f\theta_B$.

These lines are called **Bragg lines** or **Bragg contours** because they correspond to the specimen loci, which are at the exact Bragg orientation.

Note

Only the term "Bragg line" will be used in this book. The term "Bragg contour" is not really appropriate because it refers to an image observed in the image plane rather than to a diffraction pattern observed in the back focal plane of the objective lens.

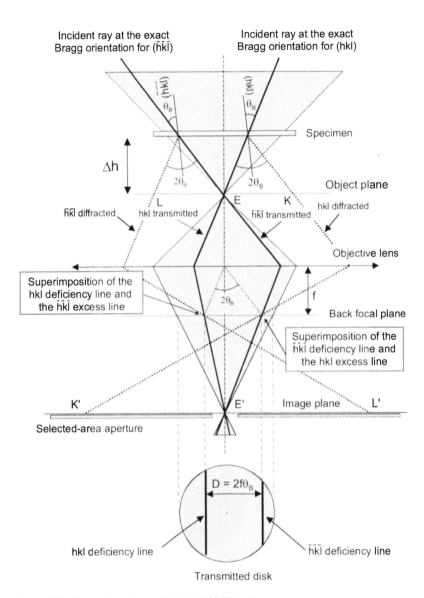

Figure VI.3 - Formation of a bright-field LACBED pattern.

Vertical cross-section through figure VI.1. Only the hkl and h̄k̄l̄ transmitted rays pass through the selected-area aperture and produce the hkl and h̄k̄l̄ deficiency lines in the transmitted disk.

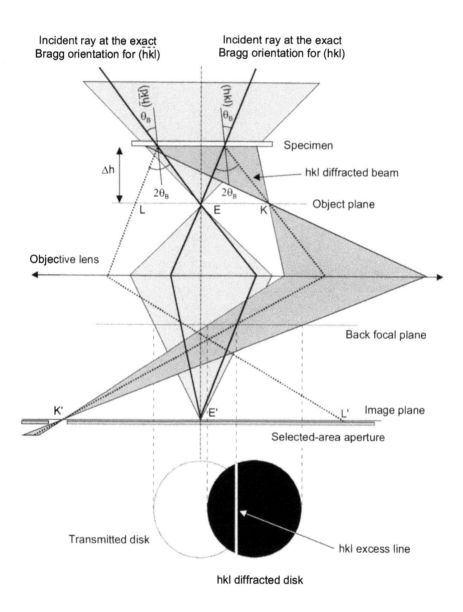

Incident ray at the exact
Bragg orientation for (h̄kl)

Incident ray at the exact
Bragg orientation for (hkl)

(h̄kl)

(hkl)

θ_B

θ_B

Specimen

Δh

hkl diffracted beam

$2\theta_B$

$2\theta_B$

Object plane

L

E

K

Objective lens

Back focal plane

K'

E'

L'

Image plane

Selected-area aperture

Transmitted disk

hkl excess line

hkl diffracted disk

Figure VI.4 - Dark-field LACBED pattern. Only the hkl diffracted rays pass through the
selected-area aperture and produce a bright hkl excess line in the hkl diffracted disk.

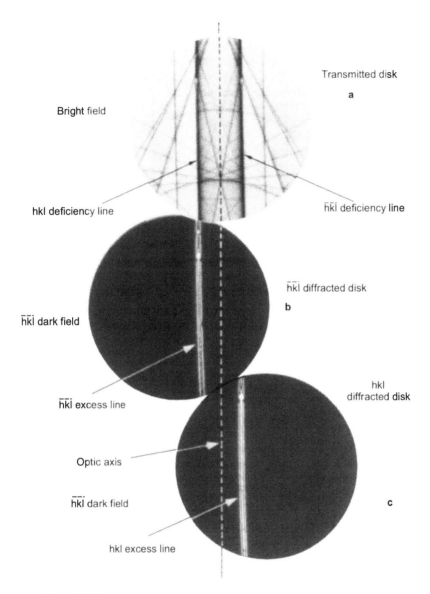

Transmitted disk

a

Bright field

hkl deficiency line

$\overline{hkl}$ deficiency line

$\overline{hkl}$ diffracted disk

b

$\overline{hkl}$ dark field

$\overline{hkl}$ excess line

hkl diffracted disk

Optic axis

$\overline{hkl}$ dark field

c

hkl excess line

Figure VI.5 - LACBED pattern.

a - Bright-field pattern. The hkl and $\overline{hkl}$ deficiency lines are present.

b - $\overline{hkl}$ dark field pattern. Only the $\overline{hkl}$ excess line is visible. This line is located at the same position as the hkl deficiency line.

c - hkl dark-field pattern. Only the hkl excess line is visible. It is located at the same position as the $\overline{hkl}$ deficiency line.

The diffraction pattern obtained under these conditions displays an excellent quality (Figure VI.2c). The quality improvement is significant compared with the corresponding Kossel pattern on figure V.7 or with the pattern obtained without a selected-area aperture on Figure VI.2b. Bragg lines are clearer, and the background is brighter. This type of pattern is called a **Bright-Field (BF)** pattern.

Taking into account the large convergence of the incident beam, the present analysis given for a single set of (hkl) lattice planes occurs simultaneously for many sets. LACBED patterns thus contain many lines. They can be sorted into pairs of parallel lines having all the properties described in paragraph V.2.3.

VI.1.1.2 - Dark-field pattern

An hkl diffracted beam can be selected with the selected-area aperture instead of the transmitted beam. An hkl diffracted disk containing its bright excess hkl line is then observed (Figure VI.4). Such a pattern is called a **Dark-Field (DF)** pattern and the quality is also excellent. Obviously, it only contains a single excess line, which might be a drawback. An example is shown on figure VI.5. We note that the excess and deficiency lines belonging to a pair are located at equivalent positions in the transmitted and diffracted disks. The hkl excess line is located at the same position as the $\overline{hkl}$ deficiency line and the $\overline{hkl}$ excess line is located at the same position as the hkl deficiency line.

VI.1.2 - [uvw] zone-axis pattern

When a [uvw] zone axis of the crystal is parallel (or almost parallel) to the axis of the convergent incident beam, the spot pattern located in the conjugate object and image planes of the objective lens is symmetrical (Figure VI.6).

In this case, the bright-field LACBED pattern displays a typical aspect connected with the zone-axis symmetry (Figures VI.7a and VI.8a). It contains several pairs of broad Bragg lines consisting of fringes that intersect in the central area of the pattern. As for Kossel patterns, these pairs of lines come from vertical lattice planes containing the [uvw] zone axis, i.e. planes in the zone with [uvw]. They correspond to the zero-order Laue zone reflections observed on CBED patterns. The pattern also displays many other sharp deficiency lines.

Figure VI.6 - Symmetrical spot pattern observed in the conjugate object and image planes of the objective lens when a zone axis is parallel to the incident beam. [001] zone-axis pattern from a silicon specimen.

Depending on the crystal under examination, the zone axis and the specimen thickness, the central area of the pattern is made up either of:

- broad black and white concentric fringes whose number changes rapidly with the specimen thickness (Figure VI.7b). As for CBED patterns, these fringes are due to two-dimensional interactions between the reflections located in the zero-order Laue zone. Since only two-dimensional interactions are involved, the whole pattern symmetry is directly related to the 10 projection diffraction groups (see reference i.3 for further explanations). For example, a 4mm symmetry is observed on figure VI.7. It is related to the $4mm1_R$ projection diffraction group.

or of

- broad black and white concentric fringes as above onto which very sharp and weak Bragg lines are superimposed (Figure VI.8b). These sharp lines are not very sensitive to thickness variations but their position depends on the accelerating voltage of the electrons. These lines are due to three-dimensional interactions and correspond to the HOLZ deficiency lines observed on CBED patterns. In this case, the pattern symmetry is related to the 31 diffraction groups [i.3].

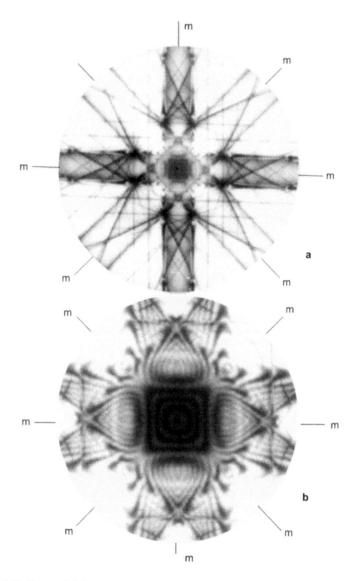

Figure VI.7 - Zone-axis LACBED pattern.
a - Bright-field <001> LACBED pattern from a silicon specimen.
b - Enlargement of the central area of the pattern. It only displays broad black and white fringes caused by two-dimensional dynamical interactions.
Both patterns display a 4mm symmetry related to the $4mm1_R$ projection diffraction group (see reference i.3 for further explanations).

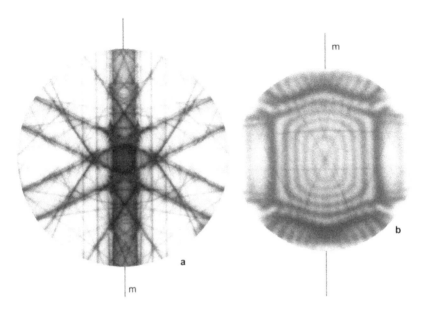

Figure VI.8 - Zone-axis LACBED pattern.
a - Bright-field <114> LACBED pattern from a silicon specimen.
b - Enlargement of the central area. It contains broad black and white concentric fringes caused by two-dimensional dynamical interactions and sharp and weak Bragg lines caused by three-dimensional dynamical interactions.
These patterns were obtained at -160°C in order to improve the visibility of the sharp Bragg lines. Both patterns display a mirror symmetry associated with the $2_R mm_R$ diffraction group. If the sharp deficiency lines are not taken into account, the pattern b displays a 2mm symmetry associated with the $2mm1_R$ projection diffraction group (see reference i.3 for further explanations).

Thus, the pattern on figure VI.8 displays m symmetry (a single mirror) connected with the $2_R mm_R$ diffraction group. We note that if the sharp HOLZ lines are not taken into account, a higher 2mm symmetry is observed for the pattern b. It is related to the $2mm1_R$ projection diffraction group.

VI.2 - Effect of the specimen height

Figure VI.9 shows the diffracted and transmitted ray-paths for various positive and negative Δh values. From that figure, we see that the spot pattern located at the image plane is magnified when the specimen is raised (Δh > 0) or lowered (Δh < 0) since the distance $D = 2\Delta h \theta_B$ separating a diffracted beam from the transmitted beam is proportional to Δh. This property is illustrated on figure VI.10.

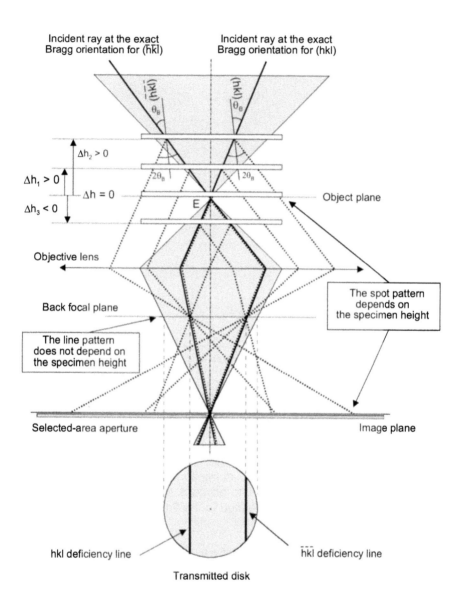

Figure VI.9 - Effect of the specimen height Δh on LACBED patterns.
The spot pattern located in the conjugate object and image planes of the objective lens depends on the specimen height Δh. On the contrary, the line pattern located in the back focal plane of the objective lens does not depend on the specimen height.

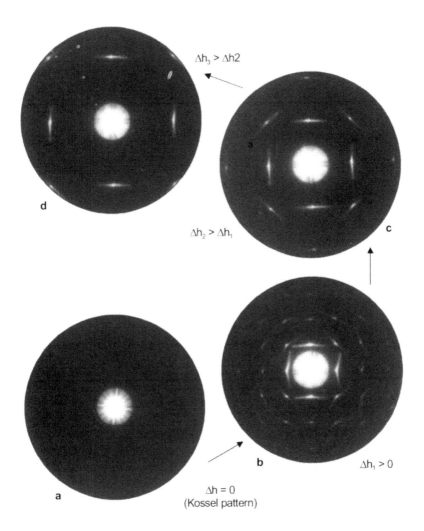

Figure VI.10 - Effect of the specimen height Δh on the spot pattern located in the conjugate object and image planes of the objective lens. The specimen height modifies the magnification of the spot pattern.
a - The specimen is located in the object plane (Δh = 0). This location corresponds to the experimental conditions used to obtain Kossel patterns. A single spot is observed.
b, c, d - The specimen is located at various positive Δh values above the object plane of the objective lens. The magnification of the spot pattern increases.
The same effects are observed for negative Δh values (specimen located below the object plane).

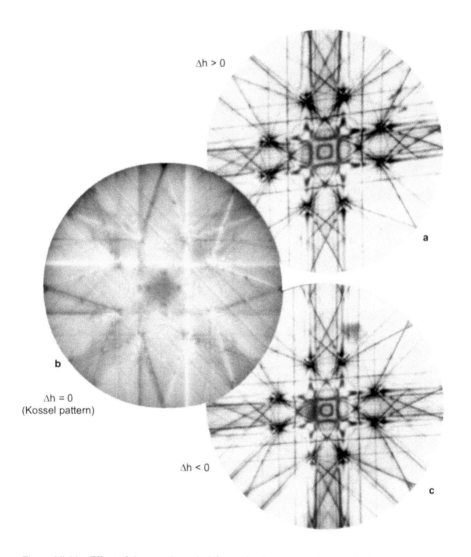

$\Delta h > 0$

a

b

$\Delta h = 0$
(Kossel pattern)

$\Delta h < 0$

c

Figure VI.11 - Effect of the specimen height on the line pattern located in the back focal plane of the objective lens. The line pattern does not depend on the specimen height.
a - The specimen is located above the object plane of the objective lens ($\Delta h > 0$).
b - The specimen is located in the object plane ($\Delta h = 0$) (Kossel pattern conditions). The excess and deficiency lines are superimposed.
c - The specimen is located below the object plane ($\Delta h < 0$).
The distortion of the LACBED patterns is due to a deformed specimen. Note that the Kossel pattern is not distorted.

Figure VI.9 also shows that the orientation of the transmitted and diffracted beams at the exit face of the specimen does not depend on the specimen height: it only depends on the Bragg angle θ_B. Since all the rays leaving the specimen with the same orientation focus at the same point in the back focal plane, **the LACBED pattern does not depend on the specimen height** (Figures VI.11).

Note that the position $\Delta h = 0$ corresponds to the experimental conditions used for Kossel patterns (Figure VI.11b). The excess and deficiency lines are then superimposed because they can no longer be isolated with the selected-area aperture.

VI.3 - Minimum specimen height Δh_{min}

In order to produce a bright- or dark-field LACBED pattern, the transmitted beam or one of the diffracted beams should be selected with the selected-area aperture. Figure VI.12 shows that this operation becomes possible as soon as the specimen height Δh is greater than a minimal value Δh_{min}, which can be calculated in the following way.

In the object plane, the separation between the diffracted and transmitted beams is $D = 2\Delta h\theta_B$ (VI.1)

In the conjugate image plane, this separation is:

$D_{image} = 2\gamma\Delta h\theta_B$ (VI.2)

γ being the magnification of the objective lens.

If d_{SA} is the diameter of the selected-area aperture, the selection of the transmitted or diffracted beams becomes possible if:

$D_{image} > d_{SA}/2$,   (VI.3)

and hence:

$\Delta h_{min} = d_{SA}/4\gamma\theta_B$ (VI.4)

Of course, in order to remove all the superimpositions of the deficiency and excess lines, this analysis must be applied to the Bragg line of the pattern corresponding to the smallest Bragg angle.

Application
What is the order of magnitude of this minimal height? For:
- an objective lens with a magnification $\gamma = 50$
- a selected-area aperture with a diameter $d_{SA} = 5\ \mu m$
- a [001] ZAP from a silicon specimen for which the smallest Bragg angle θ_B corresponds to the 220 Bragg line (at 300 kV, the $\theta_{220} = 0.293°$).
we obtain: $\Delta h_{min} = 4.9\ \mu m$

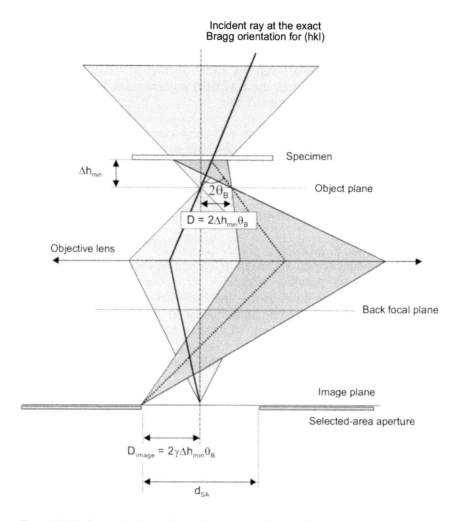

Figure VI.12 - Determination of the minimum value Δh_{min} required for the separation of the transmitted and diffracted beams with the selected-area aperture. It depends both on the diameter d_{SA} of the selected-area aperture and on the specimen height Δh.

Figure VI.13 shows a limiting case where all the reflections of the spot pattern are eliminated by the selected-area aperture except for the 220 reflections. All the lines on the corresponding LACBED pattern are deficiency lines except for the 220 lines, which correspond to the superimposition of a deficiency line and an excess line. Their contrast is very poor.

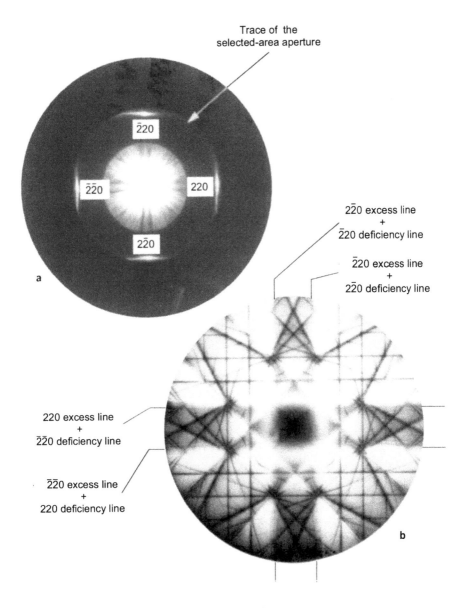

Figure VI.13 - [001] zone-axis pattern from a silicon specimen.
a - Spot pattern observed in the object plane. All the reflections are eliminated by the selected-area aperture except for the 220 reflections.
b - Corresponding LACBED pattern. All the lines are deficiency lines except for the 220 lines, which correspond to the superimposition of a deficiency line and an excess line. Their contrast is very poor.

VI.4 - Variation of the deviation parameter s

Consider Figure VI.14, where the incident convergent beam is described in terms of incident rays having all the orientations within the two limiting "-α" and "+α" orientations.

Each incident ray illuminates a different point of the specimen with a different orientation with respect to a set of (hkl) lattice planes. The ray illuminating the point A_E is at the exact Bragg orientation.

All incident rays located to the right of this point A_E have an incidence angle $\theta_I > \theta_B$. The misorientation $\Delta\theta = \theta_I - \theta_B$ with respect to the exact Bragg conditions is positive, and it becomes more important as we move away from point A_E. On the other hand, the rays located to the left of A_E have an incidence angle $\theta_I < \theta_B$. $\Delta\theta$ is now all the more negative as we move away from point A_E.

In paragraph III.1.2.3, we indicated that the misorientation can also be expressed in term of the deviation parameter s. We recall that s = 0 when the exact Bragg conditions are fulfilled; s is positive when $\Delta\theta > 0$, and negative when $\Delta\theta < 0$. Thanks to this property, the positive direction of the deviation parameter can be identified at various levels of the diffraction ray-paths using the arrows drawn on figure VI.14.

In the diffracted disk, the deviation parameter vanishes along the hkl deficiency line, and varies positively and negatively along parallel lines located on either sides of this line. The variation of s in the transmitted disk is obtained in the same way.

The analysis, carried out for the $\overline{hkl}$ transmitted and diffracted beams, leads to a deviation parameter opposite to the one related to the (hkl) planes.

Note

For a pair of hkl and $\overline{hkl}$ parallel Bragg lines located inside the transmitted disk, the arrows giving the positive direction of the deviation parameter s are opposite and point towards the outside of the transmitted disk.

We shall see that the knowledge of the sign of the deviation parameter is required for the LACBED analysis of dislocations, stacking faults and antiphase boundaries.

VI.5 - Effect of the probe size S

Until now, we have considered a point probe E. Figure VI.15 shows that with a probe size S, the specimen locus at the exact Bragg orientation is no longer a straight line but becomes a ribbon of width ΔL.

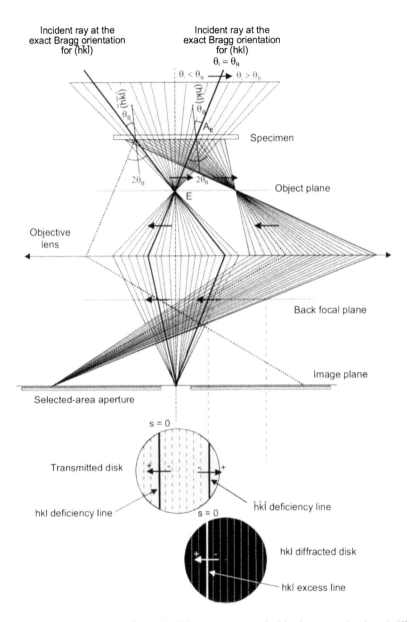

Figure VI.14 - Variation of the deviation parameter s inside the transmitted and diffracted disks. The arrows indicate the positive direction of the deviation parameter s.

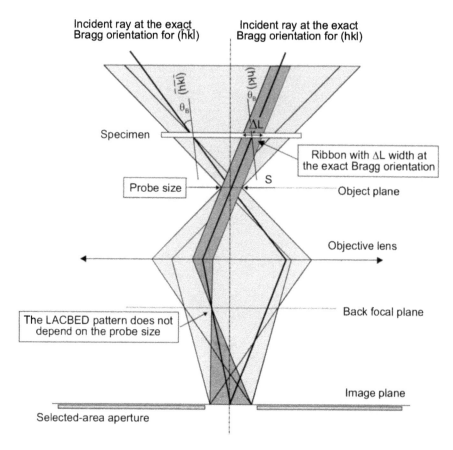

Figure VI.15 - Effect of the probe size S on LACBED patterns. A specimen ribbon of width ΔL is at the exact Bragg orientation. The probe size has no effect on the LACBED pattern.

This probe broadening has no effect on the excess and deficiency Bragg lines since all beams leaving the specimen with the same direction focus at the same point in the back focal plane of the objective lens. Therefore, the LACBED pattern does not depend on the probe size.

This property is illustrated on the experimental patterns on figure VI.16 carried out with three different probe sizes: 80 nm, 13 nm and 3 nm. As expected, no perceptible difference is observed.

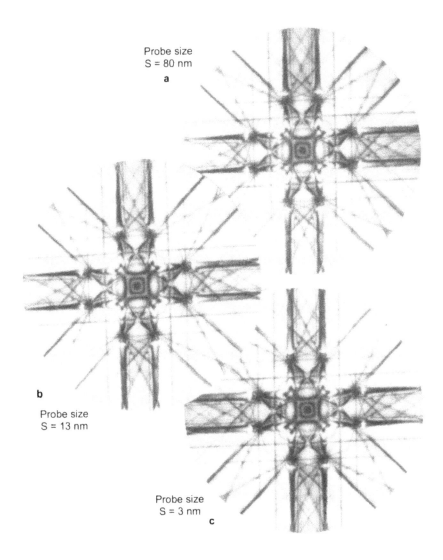

Probe size
S = 80 nm
a

b
Probe size
S = 13 nm

Probe size
S = 3 nm
c

Figure VI.16 - Effect of the probe size on [001] LACBED patterns from a silicon specimen.
a - Probe size S = 80 nm.
b - Probe size S = 13 nm.
c - Probe size S = 3 nm.
There is no perceptible difference.

VI.6 - Formation of the shadow image

Figure VI.17 shows that, in addition to the deficiency and excess lines, the LACBED pattern also contains direct-space information since the illuminated area ab of the specimen is imaged in the back focal plane of the objective lens as:
- a'b' in the transmitted disk,
- a"b" in the hkl diffracted disk,
- a'''b''' in the $\overline{hkl}$ diffracted disk.

This means that **the shadow* image of the illuminated area of the specimen is superimposed on the diffraction pattern**.

This image is called a **shadow image because it is out of focus in the back focal plane, as it will be explained in paragraph VI.6.7.*

This superimposition is clearly visible on the bright-field LACBED pattern on figure VI.18 where the image of a particle located on the specimen surface is superimposed on the Bragg lines.

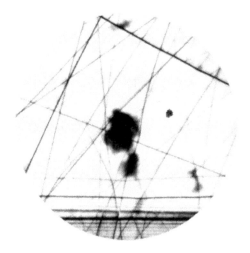

Figure VI.18 - Bright-field LACBED pattern. The shadow image of a particle located on the specimen surface is clearly visible.
Note that the particle distorts slightly some Bragg lines around it.

The shadow image superimposed on the diffraction pattern has several interesting properties, which deserve to be detailed.

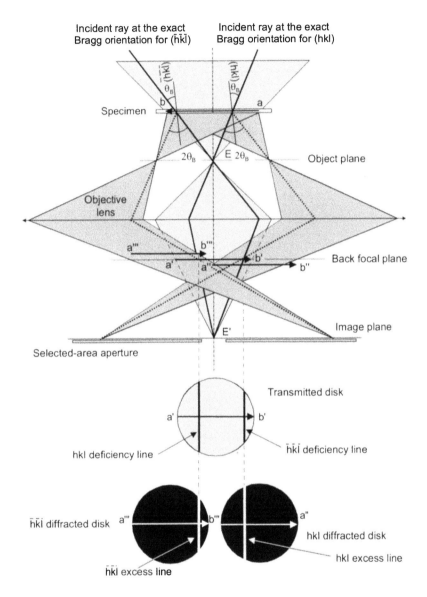

Figure VI.17 - Formation of the shadow image. The illuminated area ab of the specimen is imaged, in the back focal plane, as a'b', a"b" and a'''b''' located in the transmitted disk and in the hkl and $\overline{hkl}$ diffracted disks respectively.

VI.6.1 - Size of the illuminated area

Figure VI.19 shows that the size T of the illuminated area depends both on the specimen height Δh and on the convergence semi-angle α according to:

$$T = 2\alpha\Delta h \qquad (VI.5)$$

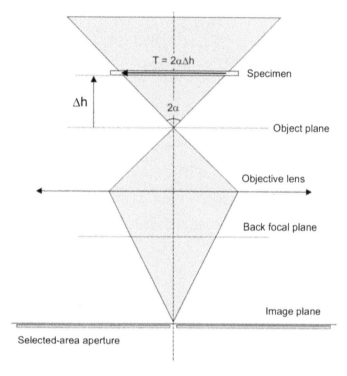

Figure VI.19 - Size of the illuminated area of the specimen. It depends both on the specimen height Δh and on the convergence semi-angle α.

VI.6.2 - Minimum size of the illuminated area

The minimum size T_{min} of the illuminated area is obtained for Δh_{min}, i.e. for the minimum Δh value required to separate the excess and deficiency lines:

$$T_{min} = 2\alpha\Delta h_{min} \qquad (VI.6)$$
$$\text{with } \Delta h_{min} = d_{SA}/4\gamma\theta_B \qquad (\text{see VI.4})$$
$$T_{min} = \alpha d_{SA}/2\gamma\theta_B \qquad (VI.7)$$

Numerical application
For:
- an incident beam with a convergence semi-angle $\alpha = 1.5°$
- $d_{SA} = 2\mu m$ and $\gamma = 50$
we obtain: $T_{min} = 0.13 \mu m$

Thus, the LACBED technique requires relatively large crystals. This is a serious limitation. In order to reduce this size, the use of a selected-area aperture with a very small diameter (from 1 to 5 μm) is recommended.

VI.6.3 - Magnification of the shadow image

The magnification γ of the shadow image a'b' (or a"b" and a'''b''') observed in back focal plane of the objective lens depends on the value of the specimen height Δh. This effect is illustrated on figure VI.20. The higher the specimen (or the lower), the larger the size of the illuminated area ab:

$$T = ab = 2\alpha\Delta h \tag{VI.8}$$

This illuminated area is imaged as a'b' in the back focal plane according to the diameter of the transmitted or diffracted disk. This diameter depends only on the convergence semi-angle α of the incident beam:

$$T' = a'b' = 2\alpha f \tag{VI.9}$$

The magnification γ is thus:

$$\gamma = T'/T = a'b'/ab = 2\alpha f / 2\alpha\Delta h \tag{VI.10}$$

Finally:

$$\gamma = f / \Delta h \tag{VI.10}$$

The magnification of the shadow image is inversely proportional to the specimen height Δh.

This effect is illustrated on the experimental patterns on figure VI.21 carried out for some positive and negative Δh values.

Note that the magnification of the image becomes infinite for $\Delta h = 0$, i.e. for the experimental conditions corresponding to a Kossel pattern.

We also recall that the line diffraction pattern is independent of the specimen height Δh.

The choice of the specimen height is particularly important, and depends greatly on the nature of the experiments being carried out. This aspect will be developed in chapter VIII, where experimental conditions are examined.

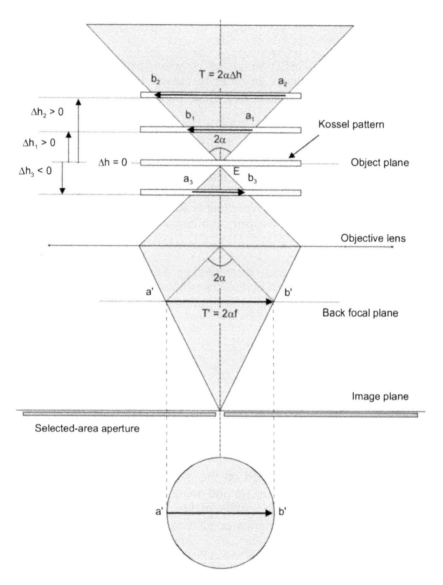

Figure VI.20 - Effect of the specimen height Δh on the magnification γ of the shadow image. The magnification is inversely proportional to Δh.

$\Delta h_3 > \Delta h_2 > \Delta h_1 > 0$ $\Delta h_2 > \Delta h_1 > 0$

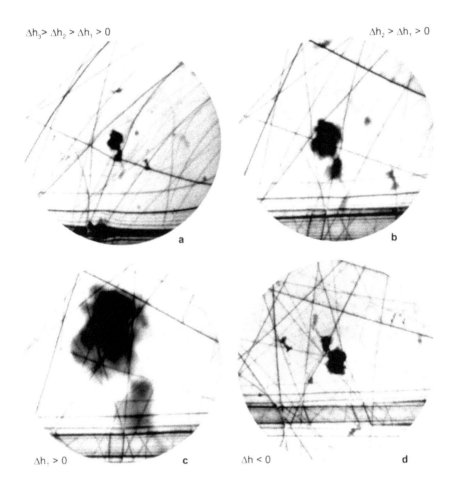

$\Delta h_1 > 0$ c $\Delta h < 0$ d

Figure VI.21 - Effect of the specimen height Δh on the magnification γ of the shadow image. The image of a particle located on the surface of the specimen is observed in the back focal plane of the objective lens for various positive and negative Δh values.

a, b, c - Positive and increasing Δh values.

d - Negative Δh value. The image is rotated by 180°.

VI.6.4 - Shift of the shadow image in bright- and dark-field LACBED patterns

Figure VI.22 shows that the shadow images A', A" and A''' of a point A of the illuminated area observed in the transmitted disk and in the hkl and $\overline{hkl}$ diffracted disks respectively are shifted with respect to one another by a distance $D = 2f\theta_B$.

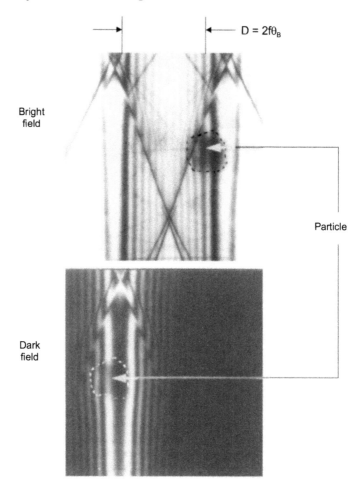

Figure VI.23 - Shift of the shadow image in bright- and dark-field LACBED patterns. The image of a particle stuck on the specimen surface is shifted by a distance $D = 2f\theta_B$ in the bright- and dark-field LACBED patterns.

This property must be taken into account during the analysis of crystal defects using bright- and dark-field LACBED patterns. Figure VI.23 gives an example of such a shift.

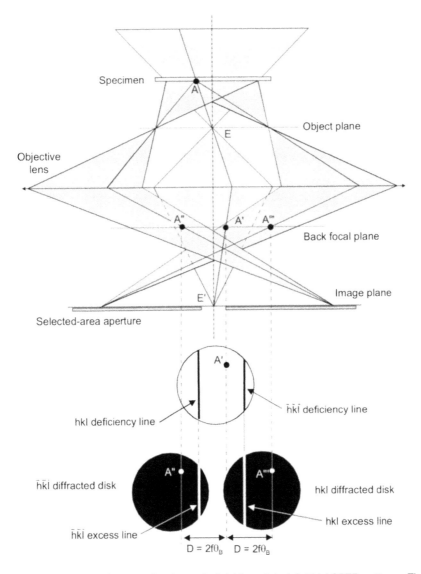

Figure VI.22 - Shifts of the shadow image in bright- and dark-field LACBED patterns. The images A', A" and A''' of a point A of the specimen are shifted by the distance $D = 2f\theta_B$ in the back focal plane of the objective lens.
Note that the images A', A" and A''' are located at equivalent positions in these three disks.

Note that the images A', A" and A"' are located at equivalent positions in the transmitted and diffracted disks.

VI.6.5 - Rotation of the shadow image

The fact of raising ($\Delta h > 0$) the specimen with respect to its normal position in the object plane results in a 180° rotation of the shadow image in the back focal plane of the objective lens (Figure VI.24). On the contrary, there is no rotation when the specimen is located below the object plane ($\Delta h > 0$). This effect is illustrated on figures VI.25a and b and it should be taken into account during the analysis of crystal defects.

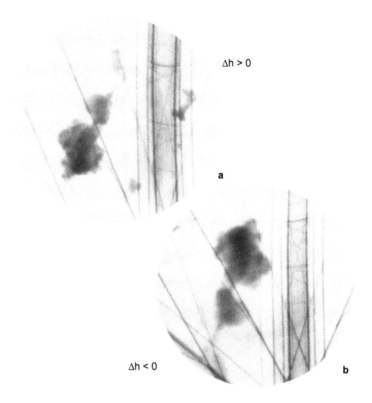

Figure VI.25 - Bright-field LACBED patterns. The shadow image of a particle stuck on the specimen surface displays a 180° rotation for positive and negative Δh values.
a - $\Delta h > 0$. The shadow image displays a 180° rotation with respect to the line pattern.
b - $\Delta h < 0$. There is no rotation of the shadow image with respect to the line pattern.

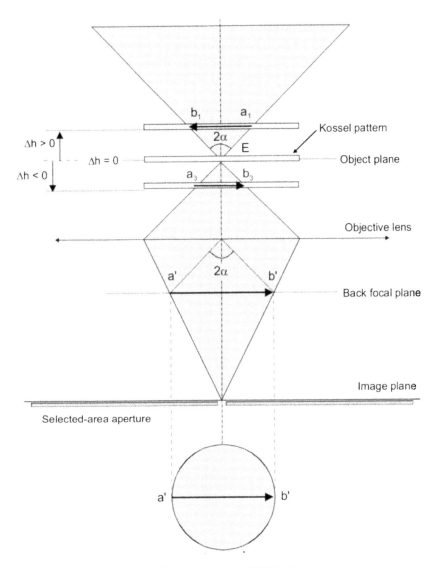

Figure VI.24 - Rotation of the shadow image on a LACBED pattern.
The image a'b' of the illuminated area ab of the specimen suffers a 180° rotation when Δh > 0
and no rotation when Δh < 0.

Note
   *A negative specimen height Δh < 0 is often preferred because it corresponds to
the absence of rotation between the illuminated area ab and its image a'b'.*

VI.6.6 - Resolution r of the shadow image

The resolution r of the shadow image depends on the probe size S. This property can be understood by imagining that this image is formed in a still camera whose lens is replaced by a small hole (pinhole camera).

It can also be demonstrated in the following way.

Consider a point A of the specimen, chosen for the sake of simplicity on the optic axis (figure VI.26). It gives a focused image A' on a plane PP' located below the focal plane. It also gives a defocused image, i.e. a disk of confusion a'b' with diameter 2r' in the back focal plane:

$$r' = a'b'/2 \qquad (VI.11)$$
$$r' = f\rho \qquad (VI.12)$$

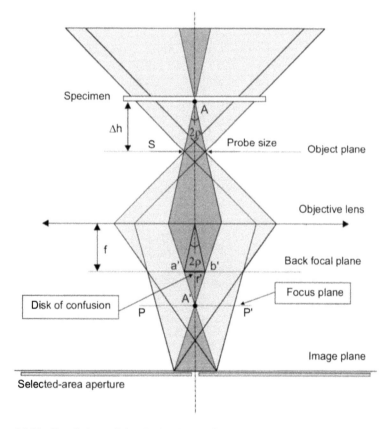

Figure VI.26 - Resolution r of the shadow image. The image of a point A of the specimen is a disk of confusion a'b' with radius r' in the back focal plane of the objective lens.

but:
$$\rho = S/2\Delta h \tag{VI.13}$$
hence,
$$r' = fS/2\Delta h \tag{VI.14}$$

In the object plane, r' appears under a disk of confusion with radius r:
$$r = r'/\gamma = fS/2\gamma\Delta h \tag{VI.15}$$

But, we indicated on paragraph VI.6.3, that:
$$\gamma = f/\Delta h \tag{seeVI.10}$$
hence,
$$r = fS\Delta h/2f\Delta h$$

Finally:
$$r = S/2 \tag{VI.16}$$

The resolution r of the shadow image is thus directly related to the probe size S. For this reason, it is advisable to use the smallest available probe sizes. This is why the nanoprobe mode is used systematically to perform LACBED experiments. The effect of three different probe sizes on the shadow image of a particle stuck on a specimen surface is illustrated on figure VI.27. With very small probe sizes, the improvement of the quality of the image is clearly observed.

VI.6.7 - Focus of the shadow image and of the line pattern

The line pattern is observed in the back focal plane of the objective lens. Figure VI.15 shows that the diffracted and transmitted beams are perfectly focused on this plane whatever the probe size is. This means that **the line pattern is in focus in the back focal plane of the objective lens**.

With regard to the image, figure VI.26 shows that it is not focused on the back focal plane, but on an image plane PP' located below the focal plane.

Thus, in a LACBED pattern, **the shadow image of the specimen is out of focus**. Using a very small probe size, S retains a reasonable image quality since the resolution r of the image is directly related to the probe size.

These properties can be illustrated by changing the adjustment of the diffraction lens of the microscope in order to observe planes located slightly above or below the back focal plane (Figure VI.28).

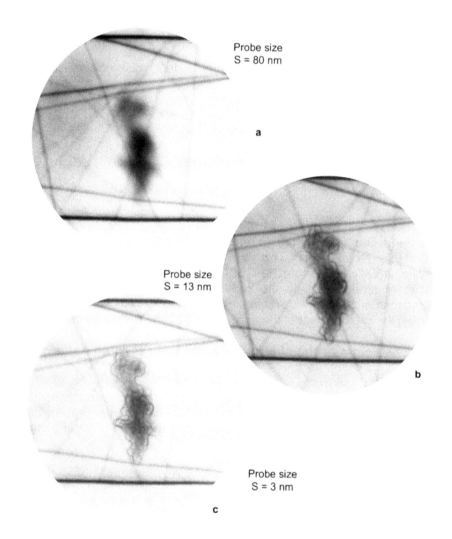

Probe size
S = 80 nm

a

Probe size
S = 13 nm

b

Probe size
S = 3 nm

c

Figure VI.27 - Aspect of the shadow images of a particle located on the specimen surface as a function of the probe size.
a - Probe size S = 80 nm.
b - Probe size S = 13 nm.
c - Probe size S = 3 nm.
A small probe size produces a dramatic improvement of the resolution of the shadow image.

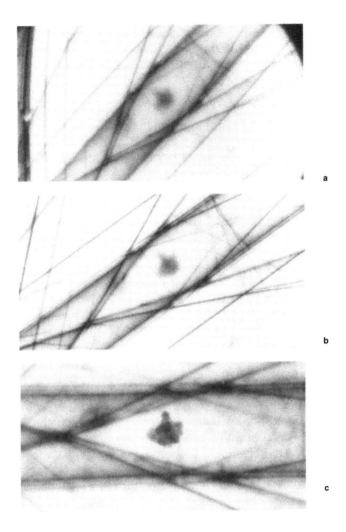

Figure VI.28 - Observation of the LACBED pattern in planes located above, in and below the back focal plane of the objective lens.
a - Plane located above the back focal plane. Neither the Bragg lines nor the image are sharp.
b - Back focal plane. The Bragg lines are sharp but the image is not.
c - Plane located below the focal plane. The lines are not sharp. The image is sharper.
Note the pattern rotation caused by the helical paths of the electrons inside the magnetic lenses.

In a plane located above the back focal plane (Figure VI.28a), neither the line pattern nor the shadow image is sharp. In a plane located below the focal plane (Figure VI.28c), the lines are not sharp but the shadow image becomes sharper because this plane is closer to the true focus plane PP'.

In chapter VIII, we shall describe the CBIM technique (Convergent-Beam Imaging) in which the image quality is favoured at the expense of the quality of the diffraction pattern. The image is then in focus whereas the excess and deficiency lines are out of focus.

VI.7 - Angular filtering of LACBED patterns

In paragraph VI.1, we reported the excellent quality of the LACBED patterns compared with the Kossel or CBED patterns. They are much sharper and show more details. The experimental CBED and LACBED patterns shown on figure VI.29 were obtained under the same experimental conditions and with the same photographic processing. The superiority of the LACBED pattern is evident. This high quality is related to the fact that the selected-area aperture acts as an efficient filter and removes most of the inelastic electrons (thermal scattering, plasmon scattering, Compton scattering...).

In order to understand this effect, consider figure VI.30 where a single incident ray is drawn. Some incident electrons undergo inelastic interactions during their passage through the specimen. They lose energy and are scattered at various angles from their initial paths. For example, consider the small and the large scattering angles ρ_1 and ρ_2 shown on figure VI.30. The electrons scattered through the large angle ρ_2 are eliminated by the selected-area aperture. The electrons scattered through ρ_1 will only be eliminated by a small aperture (drawn in dotted lines on the figure). Therefore, this filtering effect is all the more efficient as the aperture diameter is small. The quality improvement resulting from this angular filtering is visible on the LACBED patterns on figure VI.31 obtained with 10 and 50 µm selected-area apertures.

For this reason, it is advisable to use a very small selected-area aperture. Values of 1, 2 or 5 µm are strongly recommended. Thanks to these small apertures, the size of the illuminated area can also be reduced (see paragraph VI.6.2).

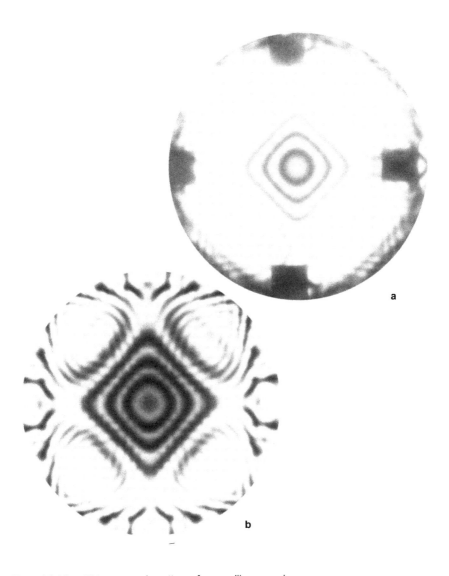

Figure VI.29 - <001> zone-axis patterns from a silicon specimen.
a - CBED pattern.
b - LACBED pattern.
The quality of the LACBED pattern is definitely higher.
The two patterns were produced with the same experimental conditions and with the same photographic processing.

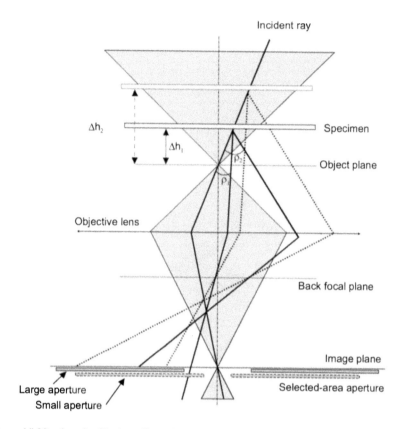

Figure VI.30 - Angular filtering effect of the selected-area aperture. The inelastic electrons are eliminated by the selected-area aperture. The smaller the selected-area aperture and/or the larger Δh (drawn in dotted lines), the larger the number of inelastic electrons removed.

Filtering is also improved by the use of large Δh values since a larger separation of the inelastic scattered rays is produced in the image plane where the selected-area aperture is located. The corresponding effect is represented in dotted lines on figure VI.30, and is illustrated on the experimental patterns of figure VI.32.

Jordan *et al.* [VI.1] analysed the effects of angular filtering in a quantitative way by varying both the size of the selected-area aperture and the specimen height Δh. They showed that most of the inelastic electrons (such as those scattered by plasmons) are eliminated. This efficient filtering constitutes a major advantage of the LACBED technique.

50 µm
aperture

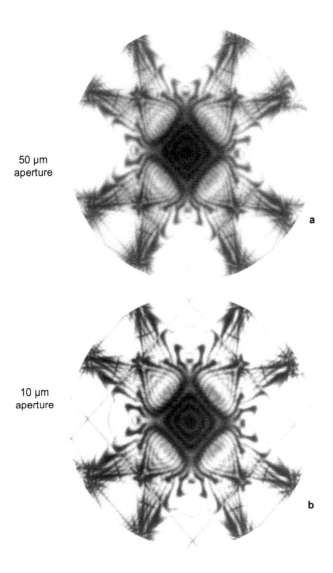

10 µm
aperture

Figure VI.31 - Effect of the size of the selected-area aperture on the angular filtering. [001]
zone axis from a silicon specimen.
a - 50 µm aperture diameter.
b -10 µm aperture diameter.
A smaller aperture removes the inelastic electrons more efficiently.

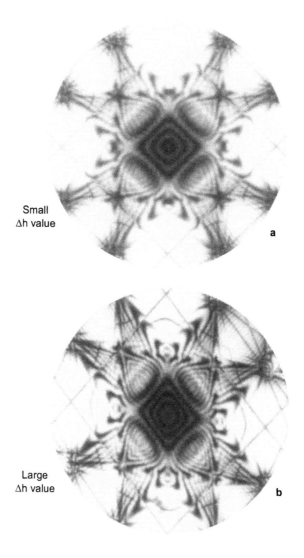

Small
Δh value

a

Large
Δh value

b

Figure VI.32 - Effects of the specimen height Δh on the angular filtering of the selected-area aperture. [001] zone axis from a silicon specimen.
a - Small specimen height Δh.
b - Large specimen height Δh.
Inelastic electrons are more efficiently removed when the specimen height Δh is greater.
The differences between the two patterns are related to the size of the illuminated area of the specimen, which changes with Δh.

Note
The probability of inelastic scattering increases with the specimen thickness; the beneficial effect of angular filtering by the selected-area aperture is only appreciable for thick specimens.

VI.8 - Effect of the temperature on LACBED patterns

The atoms in a crystal are not motionless but vibrate around their equilibrium positions all the more strongly as the temperature is high.

We can distinguish between independent vibrations constituting the Debye effect and collective vibrations resulting in the propagation of waves called phonons.

The Debye effect is most significant in electron diffraction. It corresponds to the fact that the probability of finding an atom at its equilibrium position increases to a maximum and then decreases according to a Gaussian function. The point position of the atom in the crystal is thus replaced by a Gaussian probability function.

What is the consequence of the Debye effect on the diffracted and transmitted intensities?

It acts on the atomic scattering factor f according to:

$$f_T = f_R \exp[(-B \sin^2\theta)/\lambda^2] = f_R \exp[-M] \qquad (VI.18)$$

with $M = B \sin^2\theta/\lambda^2$ $\qquad (VI.19)$

f_T is the atomic scattering factor at temperature T.

f_R is the atomic scattering factor calculated for the atom at rest, i.e. for absolute zero.

$B = 8\pi^2\sigma^2$ is the Debye-Waller factor. $\qquad (VI.20)$

σ is the average standard deviation. The Debye-Waller factor depends on the element or on the compound. Its value can be calculated from data provided in the International Tables of Crystallography [VI.2]. The calculation is relatively easy for cubic crystals with a single atom species but it becomes very complex in other cases.

The atomic scattering factor is taken into account in the structure factor F_{hkl}.

$$F_{hkl} = \Sigma f_n \exp 2\pi (hx_n + ky_n + lz_n) \qquad (VI.21)$$

x_n, y_n, z_n are the coordinates of the n-th atom in the unit cell and the summation is extended over the n atoms of the unit cell.

Within the kinematical approximation (see paragraph VII.2), the structure factor F_{hkl} is related to the diffracted intensity I_{hkl} by:

$$I_{hkl} \propto |F_{hkl}|^2 \qquad (VI.22)$$

The Debye effect results in a decrease of the diffracted intensity as exp(-2M). This decrease compensates for an increase of the background intensity according to:

$$1 - \exp(-2M) \tag{VI.22}$$

The physical meaning of the temperature effect can be understood if, by virtue to the thermal vibration, the lattice planes are considered to have a "thickness" and an interplanar spacing $d_{hkl} \pm \Delta d$. Bragg angles are also affected since they are related to these interplanar spacings by Bragg's law. They become $\theta_B \pm \Delta\theta$.

Thus, a temperature rise produces a broadening of the Bragg lines. From a qualitative point of view, these considerations lead us to expect the following improvements when a diffraction pattern is obtained at low temperature:

- the diffracted intensity is increased,
- the background intensity is reduced,
- the Bragg lines are sharper.

These improvements can be quantified by calculating B and exp(-2M). Consider the case of silicon, which is a cubic crystal containing a single atom species.

$$B = B_0 + B_T \tag{VI.24}$$

B_0 is the Debye-Waller component at 0K

B_T is the temperature dependent component.

The B_T values at 0, 93 and 273K, given in the International Tables of Crystallography [VI.2], are:

$B_0 = 0.16$ A²

$B_{93} = 0.035$ A²

$B_{273} = 0.24$ A²

The following table gives the computed values of B and exp(-2M) for various silicon reflections and for the three previous temperatures.

		0K	93K	293K
	$B = B_0 + B_T$	0.16	0.20	0.40
Reflection	θ_B	Exp(-2M)	exp(-2M)	exp(-2M)
111	0.336°	0.99	0.99	0.98
222	0.55°	0.97	0.96	0.92
400	0.77°	0.96	0.95	0.90
800	1.54°	0.84	0.81	0.65
16 00	3.08°	0.50	0.43	0.18
32 00	6.17°	0.06	0.03	0.00

Table VI.1. Calculation of B and exp(-2M) for three temperatures and various silicon reflections (specimen observed at 300 kV).

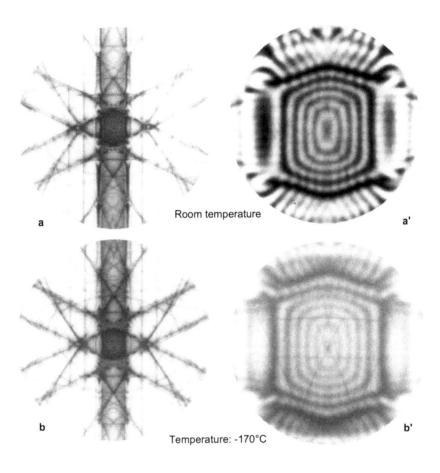

a

Room temperature

a'

b

Temperature: -170°C

b'

Figure VI.33 - Effect of the temperature on LACBED patterns. <114> zone-axis patterns from a silicon specimen.
a and a' - Pattern carried out at room temperature and enlargement of the central area.
b and b' - Pattern carried out at -170°C. Sharp and weak HOLZ lines are more visible in the central area of the pattern.

This table shows that the effect of the temperature becomes appreciable only for the 16 00 and 32 00 reflections, i.e. for large Bragg angles (about a few degrees).

This means that the use of **low temperature will only produce a significant improvement of the excess and deficiency lines corresponding to very large Bragg angles**. This is particularly the case for the sharp and weak deficiency lines present in the central area of the transmitted disk on zone-axis patterns since they correspond to the largest Bragg angles that can be found on the diffraction pattern.

This is indeed what is observed on the experimental patterns of figure VI.33 obtained at room temperature and at -170°C. Only the sharp lines present in the central area of the transmitted disk display a visible improvement. These lines are very useful for accurate measurement of the lattice parameters or local strains. For these applications, the beneficial effect of low temperature can prove to be very useful.

VI.9 - Effect of a specimen tilt

If the specimen is tilted by an angle ρ with respect to its initial position (figure VI.34), the orientation of the sets of lattice planes is modified. Consequently, the incident rays that were at the exact Bragg orientation are also modified.

On the other hand, for each set of lattice planes involved, the angle $2\theta_B$ between the hkl and $\overline{hkl}$ transmitted beams does not depend on the specimen orientation. This means that the geometry of the diffraction pattern is not modified by a specimen tilt. **The whole pattern undergoes a shift along the tilt direction**. An example is given on figure VI.35.

VI.10 - Effect of a specimen deformation

Up to now, we have made the assumption that the specimen analysed is a perfect thin single crystal with parallel faces. In reality, this is seldom the case for several reasons. The main reason is related to the size of the illuminated specimen area, which is usually large and can reach several microns. Specimen deformations are expected in such an area, as shown on figure VI.36. These deformations produce a distortion of the LACBED patterns since they modify the distances D separating the hkl and $\overline{hkl}$ deficiency lines. Specimen deformations also destroy the perfect superimposition of the hkl and $\overline{hkl}$ deficiency and excess lines in the back focal plane of the objective lens.

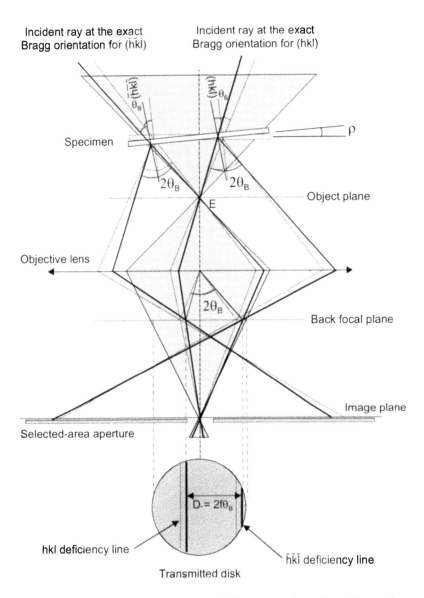

Figure VI.34 - Effect of a specimen tilt on LACBED patterns. A specimen tilt ρ produces a shift of the hkl and h̄k̄l̄ deficiency lines. The misorientation has no effect on the distance D separating these two lines.

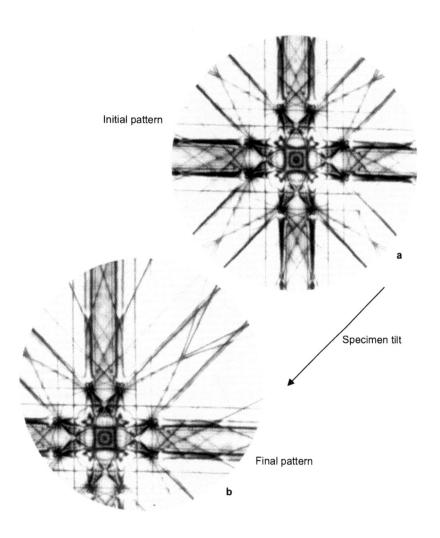

Initial pattern

Specimen tilt

Final pattern

a

b

Figure VI.35 - Effect of specimen tilt on LACBED patterns. [001] zone-axis pattern from a silicon specimen.
a - Initial pattern.
b - Pattern after a specimen tilt ρ. The whole pattern is shifted with respect to the initial pattern along the tilt direction.

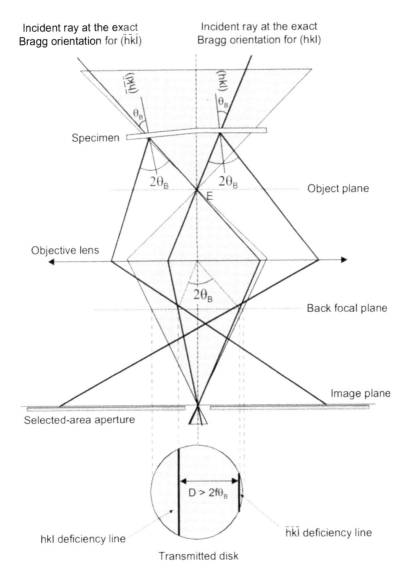

Incident ray at the exact
Bragg orientation for $(\overline{hkl})$

Incident ray at the exact
Bragg orientation for (hkl)

Specimen

Objective lens

Selected-area aperture

θ_B

$2\theta_B$

$2\theta_B$

$2\theta_B$

$D > 2f\theta_B$

hkl deficiency line

$\overline{hkl}$ deficiency line

Object plane

Back focal plane

Image plane

Transmitted disk

Figure VI.36 - Effect of a specimen deformation on LACBED patterns.
The specimen deformation produces a distortion of the pattern since the distance D separating the two hkl and $\overline{hkl}$ deficiency lines is modified. Note that the superimposition of the deficiency and excess lines is now imperfect in the back focal plane of the objective lens.

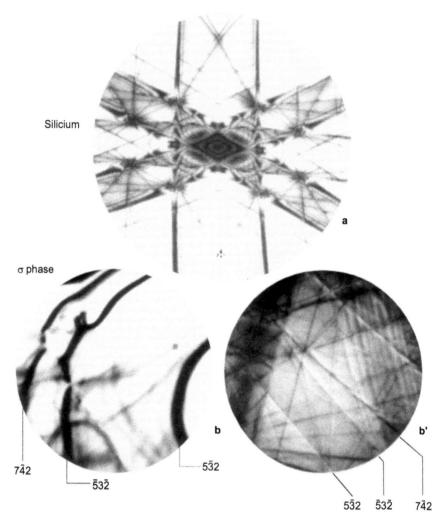

Silicium

σ phase

7$\bar{4}$2

$\bar{5}$3$\bar{2}$

5$\bar{3}$2

b

b'

5$\bar{3}$2 $\bar{5}$32   7$\bar{4}$2

Figure VI.37 - Effect of a specimen deformation on LACBED patterns.
a - <001> zone-axis pattern from a strongly deformed silicon specimen.
b - LACBED pattern from a strongly deformed σ phase specimen.
b' - Kossel pattern corresponding to the LACBED pattern b. It is undistorted and can be used to interpret the pattern b (see paragraph X).
Sigma phase specimen. Courtesy of A. Redjaïmia and N. Doukhan

Two examples of strongly distorted LACBED patterns are given on figures VI.37a and b.

The interpretation of such distorted patterns can be very difficult. In such a case, a good solution consists in raising (or lowering) the specimen height in order to switch from the distorted LACBED pattern to the corresponding undistorted Kossel pattern. An example is given on figure VI.37c.

Local strain fields can also be present around crystal defects. These strain fields distort the lattice planes and the Bragg lines but we shall see in chapter XI that they have the great advantage of producing typical effects, which can be used to characterize defects. In this case, LACBED distortions are very useful.

VI.11 - Effect of a variation of the lattice parameters

A variation of one or several lattice parameters (a, b, c, α, β and γ in the general case) produces a change of the interplanar spacings, and consequently a change of the Bragg angles θ_B since the latter are related to the interplanar spacings by Bragg's law. A shift of the excess and deficiency lines is then observed on LACBED patterns. In the case of cubic crystals, where only one lattice parameter (the lattice parameter a) is required, the angles between the lattice planes do not depend on the lattice parameter. For this reason, the lines are shifted while keeping their orientation (Figure VI.38a). For crystals belonging to the other crystal systems, the lines undergo at the same time a shift and a change of orientation (Figure VI.38b).

The sensitivity of the line shift can be estimated by differentiating Bragg's law $\lambda = 2d_{hkl}\theta_B$ with respect to θ and d_{hkl} and equating it to zero:

$$2\theta\Delta d_{hkl} + 2d_{hkl}\Delta\theta = 0 \qquad (VI.25)$$

so:

$$\Delta\theta/\theta = -\Delta d_{hkl}/d_{hkl} \qquad (VI.26)$$

In the case of a cubic crystal:

$$d_{hkl} = a/(h^2 + k^2 + l^2)^{1/2} \qquad (VI.27)$$
$$\lambda = 2d_{hkl}\theta_B = 2a/(h^2 + k^2 + l^2)^{1/2}\theta_B \qquad (VI.28)$$

finally:

$$\Delta\theta/\theta = -\Delta a/a \qquad (VI.29)$$

For a given variation Δa of the lattice parameter, the corresponding Bragg line shift $\Delta\theta$ is more significant with large Bragg angles. We recall that the lines corresponding to the highest Bragg angles are the sharp and weak lines present in the central area on zone-axis LACBED patterns (Figure VI.38).

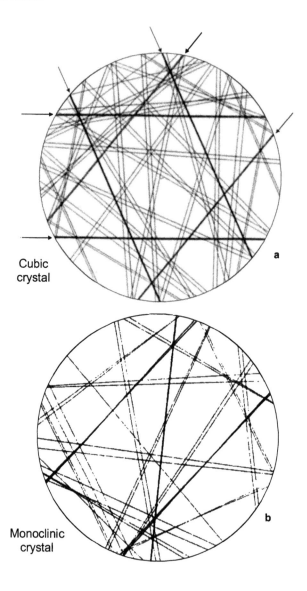

Cubic
crystal

Monoclinic
crystal

Figure VI.38. Effect of lattice parameter variations on LACBED patterns. Simulations of bright field LACBED patterns.
a - Cubic crystal. The Bragg lines are shifted while keeping their orientation. The lines marked with an arrow are almost unaffected.
b - Monoclinic crystal. The lines are shifted and their orientation is also modified.
Simulations carried out with the "Electron diffraction" software [VI.3].

On the contrary, the lines connected with lattice planes in the zone (vertical lattice planes in the microscope) have the smallest Bragg angles and are nearly unaffected; this is the case for the lines marked with an arrow on figure VI.38a.

In order to carry out accurate lattice parameter measurements, sharp HOLZ lines should be used. However, these lines often have very large extinction distances and are very weak. In this case, it is advisable to perform the experiments at low temperature in order to increase their visibility (see paragraph VI.8).

VI.12 - Effect of the accelerating voltage

A variation ΔV of the accelerating voltage results in a variation $\Delta\lambda$ of the wavelength and, thanks to Bragg's law, in a variation of the Bragg angles $\Delta\theta$. This effect is illustrated on the simulation and on the patterns of figures VI.39 and VI.40.

Cubic crystal

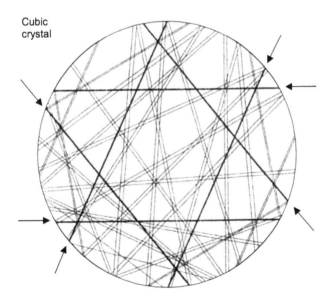

Figure VI.39 - Effect of the accelerating voltage on LACBED patterns. Simulation of the central area of the <114> zone-axis pattern from a silicon specimen observed at 120 and 123 kV. The lines are shifted without changing their orientation.
The lines corresponding to small Bragg angles (lines indicated by an arrow) are almost unaffected.

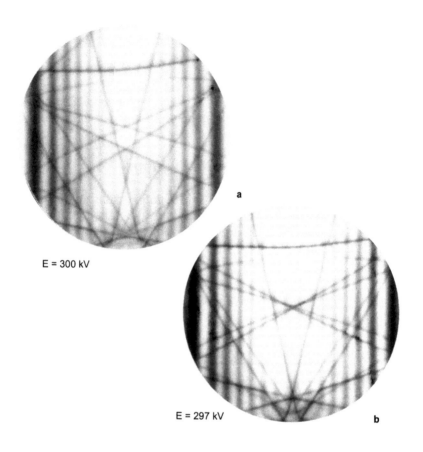

E = 300 kV

E = 297 kV

Figure VI.40 - Effect of the accelerating voltage on LACBED patterns. <114> zone-axis patterns from a silicon specimen.
a - V = 300 kV.
b - V = 297 kV.
The sharp HOLZ lines undergo a very significant shift.

Note that a variation of the accelerating voltage produces the same effect as a variation of the lattice parameters.

VI.13 - Effect of the specimen thickness

The thinner the specimen is, the broader the deficiency and excess lines are. This effect is clearly shown on figure VI.41, which give patterns for different specimen thickness.

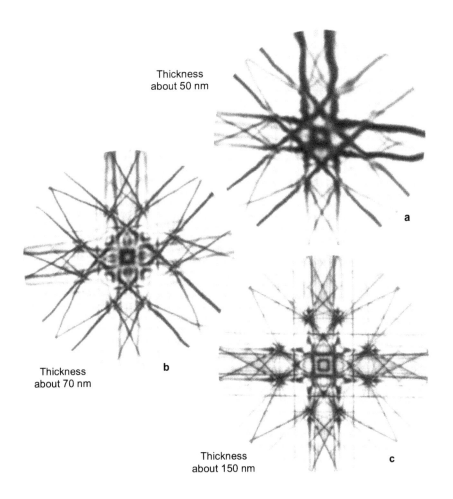

Thickness
about 50 nm

a

Thickness
about 70 nm

b

Thickness
about 150 nm

c

Figure VI.41 - Effect of a variation of the specimen thickness on LACBED patterns. <001>
zone axis-pattern from a silicon specimen.
The deficiency lines become sharper when the specimen thickness increases. The pattern
distortions are also reduced.

The best patterns are those consisting of sharp lines. They
correspond to an optimal thickness. The interpretation of this property
requires the "rocking curves" theory, which will be developed in the
following chapter.

VI.14 - Analogy with bend-contour patterns

The bend-contour patterns observed on electron micrographs from distorted specimens display a very strong analogy with the Bragg lines on LACBED patterns. For example, the similarity between the bend contour and the LACBED patterns on figure VI.42 is striking.

How are these bend-contour patterns formed?

Consider a bent specimen illuminated by a parallel incident beam (Figure VI.43). Because of the specimen curvature, the (hkl) lattice planes have a variable orientation with respect to the incident beam and some of them can be at the exact Bragg orientation. This is the case for the (hkl) lattice planes located at points A and B on figure VI.43. They give the two hkl and $\overline{hkl}$ diffracted rays. The corresponding transmitted rays are thus weakened with respect to the other transmitted rays. On a bright-field image obtained with the transmitted beam, the two images A' and B' will be darker than the images of all the other points of the specimen.

If this description is extended to a three-dimensional specimen, two black lines are observed: these lines are known as **bend contours** because they correspond to the specimen locus at the exact Bragg orientation.

Depending on the specimen curvature, this phenomenon can occur simultaneously for several sets of lattice planes and may relate to planes in the zone by giving a typical pattern. This is the case for the pattern shown on figure VI.42a.

The analogy of LACBED patterns with bend-contour patterns is significant because for both of them, the orientation of the incident beam with respect to a set of lattice planes varies at each point of the specimen.

- For LACBED patterns, the orientation of the lattice planes is the same at each point of the specimen, and it is the orientation of the incident rays that varies continuously and in a regular fashion (Figure VI.44b).

- For bend-contour patterns, the orientation of the lattice planes varies in a more or less regular fashion according to the nature of the specimen deformation whereas the orientation of the incident beam is constant (Figure VI.44a).

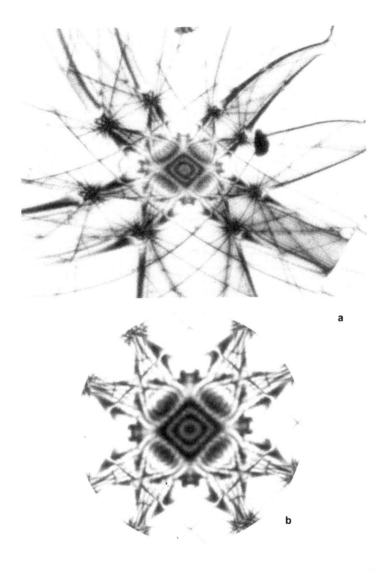

Figure VI.42 - Analogy between bend-contour and LACBED patterns from a silicon specimen.
a - Bend-contour pattern.
b - <001> LABCED pattern.
The similarity is striking.

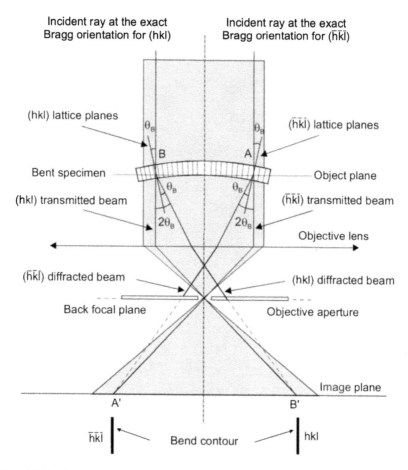

Figure VI.43 - Formation of a bend-contour pattern. At points A and B of the bent specimen, the incident electrons are at the exact Bragg orientation and produce diffracted and transmitted rays. Two dark hkl and h̄k̄l̄ bend contours are observed on a bright-field image formed with the transmitted beam (selected by the objective aperture).

The two types of patterns are however different.

LACBED patterns are observed in the diffraction mode in the back focal plane of the objective lens whereas bend-contour patterns are observed in the image mode in the image plane of the objective lens.

Strictly speaking, bend-contour patterns are not diffraction patterns. Nevertheless, they are very useful for tilting a specimen area to a specific orientation or seeking a given illumination under two-beam imaging.

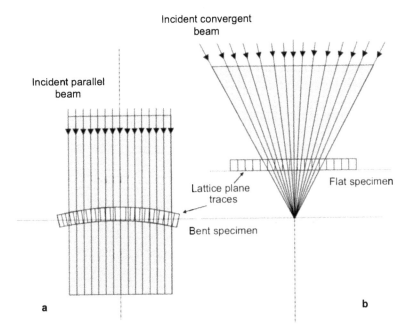

Figure VI.44 - Analogy between bend-contour patterns and LACBED patterns. The orientation of the incident rays, with respect to a set of (hkl) lattice planes, varies continuously at each point of the specimen.
a - Bend-contour pattern. The orientation of the lattice planes varies whereas the orientation of the incident rays is constant.
b - LACBED pattern. The orientation of the planes is constant whereas the orientation of the incident rays varies continuously and regularly.

It is easy to demonstrate the advantage of tLACBED patterns over bend-contour patterns.

For LACBED patterns, the convergent incident beam produces a continuous and perfectly regular variation of the incident ray orientation with respect to the lattice planes.

For the bend-contour patterns, this variation is related to the nature of the specimen curvature. Beautiful bend-contour patterns can only be obtained in very distorted specimens. In fact, they depend on the poor quality of the specimen.

On the contrary, obtaining quality LACBED patterns requires the examination of undistorted specimens. The relative misorientation mapping of the local incident rays with respect to the set of (hkl) lattice planes is important. If this mapping is regular, then quantitative measurements are possible. If not, only qualitative information is available.

This is why it is usually possible to perform measurements on LACBED patterns whereas bend-contour patterns only give nice but useless photographs. Note that bend-contour patterns have been used in real crystallography since 1973, by Steeds et al. [VI.4].

VI.15 - Analogy with Kikuchi patterns

LACBED patterns also present a very strong similarity to Kikuchi patterns (Figure VI.45).

Figure VI.45 - <001> Kikuchi pattern from a silicon specimen.

Both are line patterns produced by diffraction phenomena on sets of lattice planes, but the mechanisms of formation of the lines are different.

- For LACBED patterns, we indicated that the lines are produced by diffraction of the incident rays at the exact Bragg orientation. We recall that diffraction involves only elastic scattering.

- For Kikuchi patterns, two mechanisms are involved: a first inelastic scattering mechanism, followed by a second diffraction mechanism (elastic scattering).

Consider (Figure VI.46):

- a thick specimen illuminated by a parallel incident beam with intensity I_0.
- a set of lattice planes (hkl) whose orientation is neither equal nor close to the Bragg orientation. Consequently, no diffracted beam is produced.

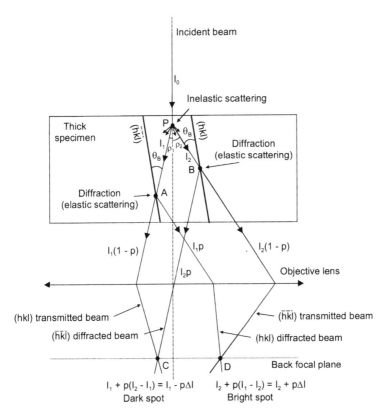

Figure VI.46 - Formation of Kikuchi lines.
Incident electrons are scattered at P. Some of them are at the exact Bragg orientation for the (hkl) lattice planes and undergo diffraction at A and B. In the back focal plane of the objective lens, they produce a dark point C and a bright point D.

However, inelastic scattering can occur at various locations of the specimen. For example at point P, electrons are scattered and change their direction. Some of them, characterized by an intensity I_1, are at the exact Bragg orientation for the set of (hkl) lattice planes and produce, at point A, an hkl diffracted ray and an hkl transmitted ray. Other electrons, with intensity I_2, are at the Bragg orientation on the other side of the set of lattice planes and give, at B, an $\overline{hkl}$ diffracted ray and an $\overline{hkl}$ transmitted ray.

The hkl transmitted ray and the $\overline{hkl}$ diffracted ray recombine at C in the back focal plane. In the same way, the $\overline{hkl}$ transmitted ray and the hkl diffracted ray recombine at D.

An estimation of the contribution of these inelastic phenomena to the intensity of the diffraction pattern at points C and D is obtained in the following manner. On figure VI.46, the intensity I_2 is lower than I_1 since the scattering angle ρ_2 is larger than ρ_1 (the probability of inelastic scattering decreases with the scattering angle ρ).

We characterize the diffracted efficiency of the set of (hkl) lattice planes by a parameter p, which represents the ratio of the diffracted intensity to the incident intensity. At point A, the incident ray with intensity I_1 produces an hkl diffracted beam with intensity $I_1 p$ and a hkl transmitted ray with intensity $I_1(1 - p)$.

In the same way, at B, the diffracted ray has an intensity $I_2 p$, and the transmitted ray, an intensity $I_2(1 - p)$.

In the back focal plane, the intensity at point C is:

$I_c = I_2 p + I_1(1 - p) = I_1 + p(I_2 - I_1)$ (VI.30)

Since $I_1 > I_2$:

$I_c = I_1 - p\Delta I$ (VI.31)

with $\Delta I = I_1 - I_2$ (VI.32)

At point D:

$I_D = I_1 p + I_2(1 - p) = I_2 + p(I_1 - I_2)$ (VI.33)

This time:

$I_D = I_2 + p\Delta I$ (VI.34)

In the absence of diffraction, the scattered intensities obtained at points C and D (background intensity) would be respectively I_1 and I_2. Consequently, the point C is darker than the background while the point D is brighter.

If this analysis is extended to the three-dimensional case (Figure VI.47), the incident, diffracted and transmitted rays are located on the Kossel cones described in chapter I. The intersection of these cones with the back focal plane (or with the screen) gives two hyperbolae: a dark deficiency hyperbola and a bright excess one. Since the Bragg angles are small, these two hyperbolae can be considered as two parallel straight lines separated by a distance $D = 2f\theta_B$. The trace of the set of (hkl) lattice planes is located halfway between these two lines. The line near the optic axis is always the dark line and it is called the **Kikuchi deficiency line**. The other bright line is called the **Kikuchi excess line** (Figure VI.48a).

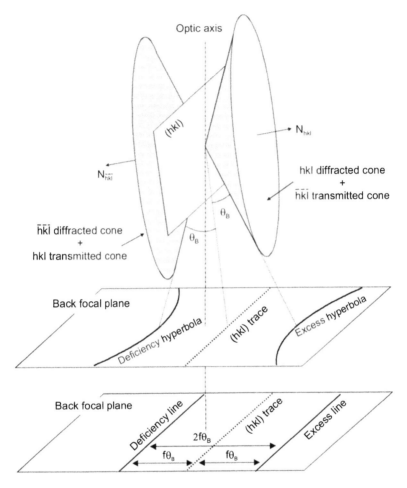

Figure VI.47 - Formation of a Kikuchi pattern.

The intersection of the hkl and $\overline{hkl}$ transmitted and diffracted Kossel cones with the back focal plane (or the screen) produces two excess and deficiency Kikuchi hyperbolae. Taking into account the low value of the Bragg angles, these two hyperbolae can be regarded as straight parallel lines. The line near the optic axis is the dark deficiency Kikuchi line.

Note

This simple analysis only gives some qualitative aspects of the Kikuchi lines. In particular, it is no longer valid in the case of a set of vertical (hkl) lattice planes where $I_1 = I_2$. The present analysis then predicts the absence of line contrast, in disagreement with experimental observations showing a homogeneous intensity stripe (Figure VI.48b). A full interpretation of Kikuchi lines is more complex and requires the use of the dynamical theory.

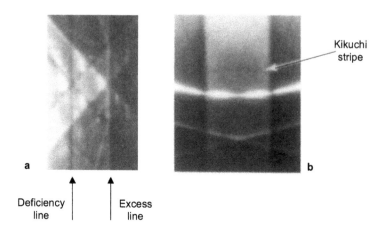

Figure VI.48 - Kikuchi lines.
a - Pair of excess and deficiency Kikuchi lines.
b - Pair of Kikuchi lines corresponding to a set of vertical lattice planes. A Kikuchi stripe is observed.

This interpretation also applies to a convergent incident beam. It is quite similar, at least in its final part, to the explanation of given the formation of the deficiency and excess lines on CBED patterns. The first inelastic scattering stage is different and exists only with Kikuchi lines. For this reason, the excess and deficiency lines of CBED patterns are sometimes called pseudo-Kikuchi lines.

VI.15.1 - Indices of the Kikuchi lines

For each Kikuchi line, the hkl diffracted Kossel cone and the $\overline{hkl}$ transmitted Kossel cone are involved. In this book, we will assign the indices of the hkl diffracted cone to the Kikuchi lines.

VI.15.2 - Properties of Kikuchi lines

Kikuchi lines display some interesting features.
Their visibility increases with the specimen thickness. The lines become clearly visible as soon as the latter is larger than a hundred nanometres. This effect is related to the probability of inelastic scattering, which increases with the specimen thickness.
The quality of the Kikuchi lines greatly depends on the crystal perfection. A specimen containing crystal defects generally displays broad and fuzzy Kikuchi lines. A small probe size also increases the

visibility of Kikuchi lines since the probability of finding imperfections in the diffracted area is strongly reduced. For this reason, Kikuchi lines are much more visible on CBED patterns than on selected-area diffraction patterns.

Inelastic interactions produce a divergence of the incident beam, the semi-angle α of which can reach several degrees (Figure VI.49a). With 300 kV incident electrons, the beam divergence is of the order of ten degrees, which is a very large value compared to the Bragg angles. It is larger than those obtained with LACBED patterns. If physical phenomena are not taken into account, a divergent beam with semi-angle α resulting from interaction with the specimen has the same effect as a convergent incident beam α (Figure VI.49b). This feature explains the great similarity between Kikuchi, Kossel and LACBED patterns.

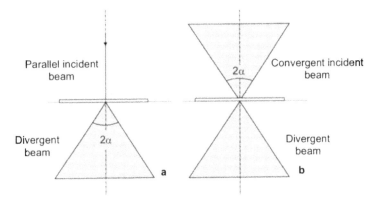

Figure VI.49 - Analogy between Kikuchi and Kossel patterns.
a - Kikuchi pattern. Inelastic scattering produces a divergence α of the incident beam.
b - Kossel pattern. The incident beam has a convergence of semi-angle α.

Like the excess and deficiency lines present on CBED and LACBED patterns, Kikuchi lines also arise from lattice planes. Consequently, they are "attached" to the specimen and undergo a shift when the specimen is tilted. On the other hand, they remain stationnary when the orientation of the incident beam is modified.

In chapter IV, we indicated that diffraction produces the information inside the transmitted and diffracted disks of a CBED pattern. All the information outside the disks comes only from inelastic scattering phenomena. All the lines observed outside the disks are, thus, Kikuchi lines (Figure VI.50). This does not necessarily mean that there are no Kikuchi lines inside the disks.

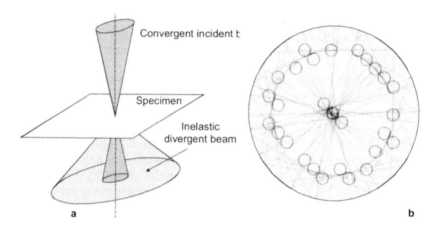

Figure VI.50 - CBED and Kikuchi patterns.
a - The information inside the disks comes from the incident rays located in the convergent incident beam. The information present outside the disks (background and Kikuchi lines) comes from the inelastic divergent rays formed during the crossing of the specimen.
b - Simulation of the zero-order Laue zone of a CBED pattern and of its Kikuchi lines. The excess and deficiency lines (bold lines) are inside the transmitted and diffracted disks. Kikuchi lines (sharp lines) are outside the disks in the continuation of the excess and deficiency lines.
Simulation carried out with the "Electron Diffraction" software [VI.3].

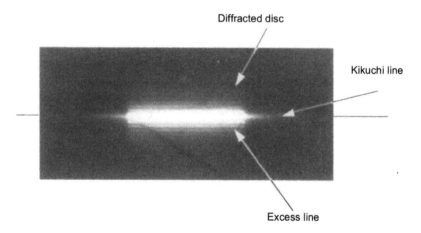

Figure VI.51 - The Kikuchi lines are in the continuation of the excess and deficiency lines.

Inelastic scattering occurs with very little energy loss. Thus Kikuchi lines are in the continuation of the excess and deficiency lines found inside the disks (Figure VI.51).

In the case of a vertical set of (hkl) lattice planes, the trace of the set pass through the optic axis (Figure VI.52a) and the hkl and $\overline{hkl}$ reflections are located at a distance $2f\theta_B$ from centre of the pattern (see paragraphs III.1.3.2 and IV.1.3). The corresponding Kikuchi lines are located halfway between the transmitted beam and the centres of the two diffracted disks. They pass through the values $s = 0$.

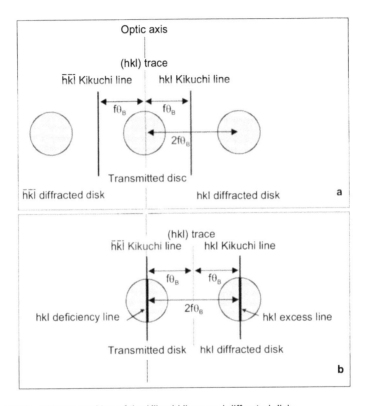

Figure VI.52 - Relative position of the Kikuchi lines and diffracted disks.
a - The set of (hkl) lattice planes is vertical. The Kikuchi lines are located halfway between the transmitted disk and the hkl and $\overline{hkl}$ diffracted disks.
b - The set of (hkl) lattice planes is at the exact Bragg orientation. The Kikuchi lines are in the continuation of the hkl excess and deficiency lines. The later pass through the centres of the transmitted and diffracted disks.

In the case of a set of (hkl) lattice planes at the exact Bragg orientation, the trace of the plane is located at a distance $f\theta_B$ from the centre of the pattern. Then, the hkl Kikuchi line pass through the centre of the diffracted disk and the other line through the centre of the transmitted disk. The Kikuchi lines are in the continuation of the excess and deficiency lines (Figure VI.52b). This configuration is very useful for the examination of defects.

The intensity of the Kikuchi lines is generally much lower than those of the excess and deficiency lines. Kikuchi lines have all the properties of the excess and deficiency lines studied in the paragraph V.2.3. In particular, the intersections of Kikuchi lines correspond to zone axes.

VI.15.3 - Interest of Kikuchi lines

Kikuchi lines are very useful for tilting a crystal to a very accurate orientation with respect to the electron beam. The large divergence associated with the Kikuchi patterns (about 10° in semi-angle !) means that zones axes are easy to identify. By tilting the specimen, Kikuchi maps can be obtained: they are very useful for the interpretation of LACBED patterns. An example will be given in chapter X.

CHAPTER VII
Diffracted and transmitted intensities

VII.1 - Generalities

In the previous chapters, we have concentrated on the geometrical aspects of the diffraction patterns. The full analysis of the patterns requires an examination of the transmitted and diffracted intensities.

Diffraction patterns obtained with a parallel incident beam consist of transmitted and diffracted spots. This means that each hkl diffracted spot can be characterized by a unique and well-defined value of the deviation parameter s characterizing the deviation from the exact Bragg orientation (Figure VII.1a).

Convergent-beam diffraction patterns consist of disks containing excess and deficiency lines. They are more complex than spot patterns since a variation of s is observed inside each hkl diffracted disk. This variation occurs along lines parallel to the excess line obtained for s = 0 (Figure VII.1b), this line being perpendicular to the vector $\mathbf{g}_{hkl}$. The same variation occurs for the deficiency line located inside the transmitted disk. The explanation of this property has been given in chapter IV.

In both cases, we can expect that the hkl diffracted intensity will depend on the following three parameters:
- the nature of the set of (hkl) diffracting planes,
- the value of the deviation parameter s,
- the specimen thickness t.

This intensity also depends, in a very significant way, on the **interactions** of the diffracted beams with the transmitted beam, as well as on the interactions between the various diffracted beams.

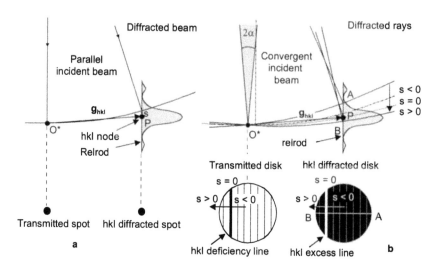

Figure VII.1 - Ewald sphere construction for parallel and convergent incident beams.
a - Parallel incident beam. A single and definite value of the deviation parameter s is associated with the hkl diffracted spot.
b - Convergent incident beam. The deviation parameter s varies continuously in the hkl diffracted disk along lines parallel to the hkl excess line obtained for s = 0. The same variation occurs in the transmitted disk.

Two theories are available to calculate the diffracted intensities: the kinematical and dynamical theories. These two theories can be applied to diffraction patterns obtained under the two-beam or the many-beam conditions. We concentrate mainly on the two-beam case since its mathematical formalism is considerably simpler than the many-beam formalism. In addition, it is often encountered on LACBED patterns.

VII.2 - Kinematical theory under two-beam conditions

This theory, initially developed for X-ray diffraction, is based on the following assumption: the single diffracted beam has a very small intensity compared to the intensity of the incident beam. This assumption implies that the interaction between the diffracted beam and the transmitted beam is negligible. It also means that diffraction only occurs once during the passage through the specimen.

In terms of wave vectors, the kinematical behaviour can be schematised by $k_0 \rightarrow k$ for the diffracted beam and $k_0 \rightarrow k_0$ for the transmitted beam (Figure VII.2a).

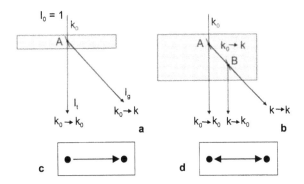

Figure VII.2 - Description of the two-beam conditions in terms of wave vectors.
a - Kinematical behaviour. Diffraction occurs only once (for example at point A) during the passage through the specimen giving a diffracted beam **k** and a transmitted beam **k₀**.
b - Dynamical behaviour. The incident beam undergoes a first diffraction (at A) and gives a transmitted beam **k₀** and a diffracted beam **k**. The diffracted beam **k** undergoes a second diffraction (at B) and produces a transmitted beam **k** and a diffracted beam **k₀**. These phenomena can occur several times inside the specimen.

In this theory, the intensity I_g of a diffracted beam produced by an incident beam with intensity $I_o = 1$ is given by the relation:

$$I_g = |\Phi_g|^2 = [1/(s\xi_g)^2] \sin^2(\pi ts) \qquad (VII.1)$$

where Φ_g is the diffracted wave amplitude and ξ_g is the extinction distance. The latter is a dynamical quantity whose meaning will be given in the following paragraph. It is related to the modulus of the structure factor $|F_g|$ by:

$$\xi_g = [\pi v_0 \cos\theta_B] / \lambda |F_g| \qquad (VII.2)$$

v_0 being the volume of the unit cell.

Equation VII.1 indicates that the diffracted intensity varies in a periodic manner as a function of both the deviation parameter s and the specimen thickness t.

VII.2.1 - Effect of the deviation parameter s on the diffracted intensity I_g

The effect of the deviation parameter s on the diffracted intensity I_g is clearly shown on the curve $I_g = f(s)$, drawn for constant values of t and ξ_g. This so-called **rocking curve** can be obtained experimentally by measuring the diffracted intensity from a flat specimen whose set of (hkl) lattice planes is rocked on both sides of the exact Bragg orientation (Figure VII.3a).

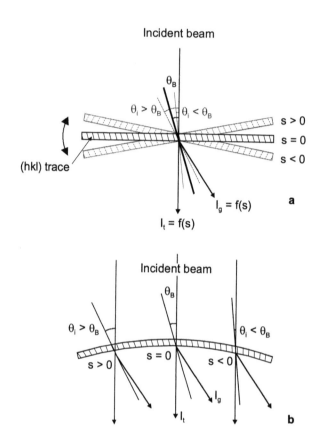

Figure VII.3 - Kinematical theory. Rocking curve $I_g = f(s)$ under two-beam conditions.
a - The rocking curve can be obtained by rocking the specimen on both sides of the exact Bragg orientation.
b - The rocking curve can also be obtained from a bent specimen illuminated by a parallel incident beam leading to the formation of bend contours (see paragraph VI.13).

The rocking curve can also be obtained from a bent specimen of constant thickness t illuminated by a parallel incident beam (Figure VII.3b). We have already described this configuration in the paragraph VI.13 and shown that it produces bend contours.

The first term of equation (VII.1):
$$1/(s\xi_g)^2 \qquad\qquad (VII.3)$$
is the envelope function of the rocking curve and the second term:
$$\sin^2(\pi t s) \qquad\qquad (VII.4)$$
its oscillations (Figure VII.4).

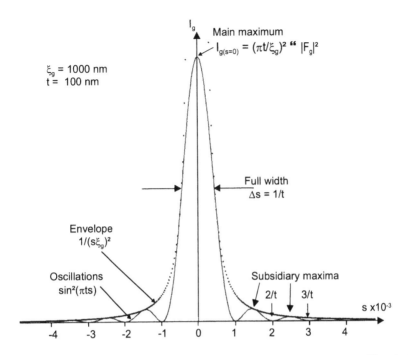

Figure VII.4 - Rocking curve $I_g = f(s)$ in the two-beam kinematical conditions. The single main maximum, obtained at $s = 0$, is proportional to the squared modulus of the structure factor $|F_g|^2$. The full width at half-height is $\Delta s = 1/t$.

The rocking curve displays a single main maximum at $s = 0$, and several weak subsidiary maxima. The diffracted intensity cancels out for $s = n/t$ where n s an integer.   (VII.5)

Taking into account the weak intensity of the subsidiary maxima, we can consider that the full width at half-height is:

$$\Delta s = 1/t \qquad\qquad\qquad\qquad (VII.6)$$

Moreover, for $s = 0$:

$$I_g = (\pi t/\xi_g)^2 = (t\lambda\,|F_g|\,/\,v_0\cos\theta_B)^2 \propto |F_g|^2 \qquad (VII.7)$$

This means that the rocking curve maximum is directly proportional to the squared modulus of the structure factor $|F_g|^2$. This value, therefore, gives access to structural determinations.

VII.2.2 - Effect of the thickness t on the diffracted intensity I_g

The diffracted intensity also varies in a periodic way with the specimen thickness t. This property is shown on the curve $I_g = f(t)$ drawn for constant s and ξ_g values (Figure VII.5a).

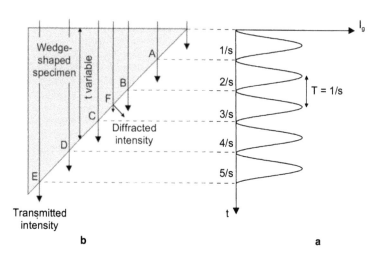

Figure VII.5 - Two-beam kinematical theory. Effect of the specimen thickness on the diffracted intensity I_g.
a - The curve $I_g = f(t)$ falls to zero for $t = k/s$ where k is an integer.
b - The curve $I_g = f(t)$ can be obtained from a wedge-shaped specimen illuminated by a parallel incident beam. At points A, B, C, D and E of the exit face of this specimen, the diffracted intensity vanishes and the transmitted intensity has a maximum value. Everywhere else, for example at the point F, the diffracted intensity is not zero and the transmitted intensity is weakened. The image obtained either from the transmitted beam (bright-field image) or from the diffracted beam (dark-field image), displays thickness fringes.

This curve can be obtained experimentally from a wedge-shaped crystal (its thickness t is variable but the value of the deviation parameter s is constant), illuminated by a parallel incident beam (Figure VII.5b). The expression:

$$I_g = [1/(s\xi_g)^2] \sin^2(\pi ts) \qquad (VII.1)$$

can be rewritten:

$$I_g = K \sin^2(\pi ts) \qquad (VII.8)$$

Its value cancels out for $\sin(\pi ts) = 0$, i.e. for thicknesses:

$$t = k/s, \text{ where k is an integer,} \qquad (VII.9)$$

therefore, for a period of $T = 1/s$ $\qquad (VII.10)$

The vanishing of the diffracted intensity for $t = k/s$ explains the presence of fringes known as **thickness fringes**, observed on bright- and dark-field images of wedge-shaped specimens.

This kinematical theory presents many imperfections. The two most significant ones concern the exact Bragg conditions.

Thus, for s = 0, we have:

$I_g = (\pi t/\xi_g)^2$ (VII.11)

If $t > \xi_g/\pi$, the diffracted intensity I_g becomes higher than 1, (i.e. higher than the incident intensity I_0), which is obviously impossible.

Moreover, for s = 0, the period T = 1/s of the thickness fringes becomes infinite. This means that the fringes should disappear when the exact Bragg conditions are satisfied. Experimentally, they remain present and have a well-defined periodicity. It is under these conditions that they are the most visible.

VII.3 - Dynamical theory under two-beam conditions

This theory is valid when the diffracted beam has a high intensity with respect to the transmitted beam. It mostly arises in electron diffraction, because electron-crystal interactions are particularly strong (they are much stronger than the X-ray-crystal interactions).

Under these conditions, the strong diffracted beam ($\mathbf{k_0} \to \mathbf{k}$) can behave like an incident beam and undergo a second diffraction process by producing a transmitted beam ($\mathbf{k} \to \mathbf{k}$) in the direction of the diffracted beam and a diffracted beam ($\mathbf{k} \to \mathbf{k_0}$) in the direction of the transmitted beam (Figure VII.2b). This phenomenon can occur several times during the passage through the specimen and becomes more likely with increasing specimen thickness. These interactions will modify the transmitted and diffracted intensities strongly. Note that these dynamical phenomena do not produce new diffracted beam directions: they do not produce new reflections or new diffracted disks.

In this dynamical theory, the diffracted intensity I_g is given by the relation:

$I_g = |\Phi_g|^2 = \sin^2\beta \, \sin^2(\pi\Delta kt)$ [VII.1] (VII.12)

β is a dimensionless extinction distance related to s and ξ_g by:

$\cot\beta = s\xi_g$ (VII.13)

and $\Delta k = [1 + (s\xi_g)^2]^{1/2} \, \xi_g^{-1}$ (VII.14)

The diffracted intensity can also be written in the same form as the kinematical expression (VII.1):

$I_g = |\Phi_g|^2 = [1/(s'\xi_g)^2] \sin^2(\pi ts')$ (VII.15)

The deviation parameter s is simply replaced by an **effective deviation parameter** s,' where $s' = (s^2 + 1/\xi_g^2)^{1/2}$ (VII.16)

VII.3.1 - Effect of the deviation parameter s on the diffracted intensity I_g

The appearance of the dynamical rocking curve is given on figure VII.6.

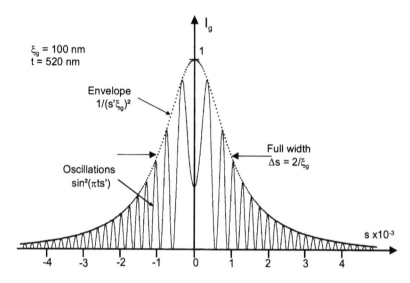

Figure VII.6 - Rocking curve $I_g = f(s)$ under two-beam dynamical conditions. This curve exhibits several maxima. The diffracted intensity does not have a maximum value for s = 0. The full width at half-height is $\Delta s = 2/\xi_g$

As in the kinematical case, the first term $1/(s'\xi_g)^2$ represents the envelope, and the second one, $\sin^2(\pi ts')$, the oscillations of the curve.

The curve displays numerous very significant maxima. The diffracted intensity does not have a maximum value at s = 0. A maximum intensity is only obtained when $t = (2n + 1)\,\xi_g / 2$, where n is an integer (Figure VII.7a).

It is even possible to obtain a zero diffracted intensity, at s = 0; this occurs when $t = n\xi_g$ (Figure VII.7b). The meaning of the extinction distance ξ_g appears clearly. **It corresponds to the smallest specimen thickness for which the diffracted intensity is cancelled**.

The full width at half-height of the envelope function of the rocking curve is related to the extinction distance by:

$$\Delta s = 2/\xi_g. \tag{VII.17}$$

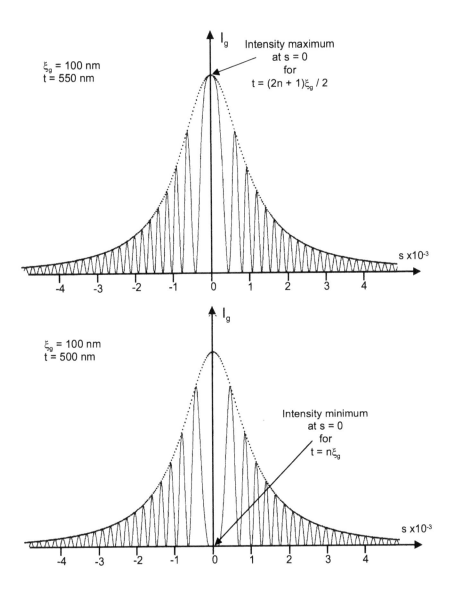

Figure VII.7 - Rocking curve $I_g = f(s)$ under two-beam dynamical conditions.
a - A maximum intensity is obtained at s = 0 when $t = (2n + 1)\,\xi_g / 2$.
b - The diffracted intensity cancels out at s = 0 when the specimen thickness t is an integer number of extinction distances ξ_g.

VII.3.2 - Effect of the thickness t on the diffracted intensity I_g

For a constant value of s', I_g can be rewritten:

$I_g = K' \sin^2(\pi t s')$ (VII.18)

I_g cancels out for:

$t = k/s'$, where k is an integer   (VII.19)

and, hence, for a period $T = 1/s'$ (VII.20)

For s = 0:

$s' = 1/\xi_g$ (VII.21)

and $t = \xi_g$ (VII.22)

The periodicity of the intensity is equal to the extinction distance ξ_g. The first imperfection of the kinematical theory is thus removed in the dynamical theory (Figure VII.8).

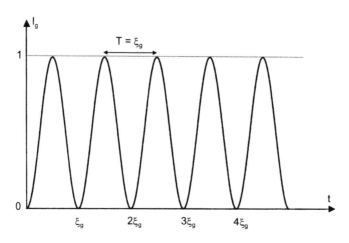

Figure VII.8 - Two-beam dynamical theory. Effect of the specimen thickness t on the diffracted intensity I_g. For s = 0, the diffracted intensity I_g cancels out when $t = k\xi_g$, where k is an integer.

VII.3.3 - Complementarity of the transmitted and diffracted intensities

For s = 0, the diffracted intensity can be written:

$I_g = \sin^2(\pi t/\xi_g)$ (VII.23)

It cancels out for $t = k\xi_g$ as mentioned above. The corresponding transmitted intensity I_t is:

$I_t = \cos^2(\pi t/\xi_g)$ (VII.24)

And, so, $I_t + I_g = 1$ (VII.25)

The transmitted and the diffracted intensities are thus complementary. This effect, which remains valid for other s values, is illustrated on figure VII.9. It means that the diffracted and the transmitted waves oscillate in opposite phase according to the specimen thickness.

VII.3.4 - Effect of absorption on the transmitted and diffracted intensities.

In the above section we have neglected all absorption of the electrons by the specimen. When absorption is taken into account, the expression (VII.12) becomes:

$$I_g = |\Phi_g|^2 = [\sin^2\beta \, (\sin^2(\pi\Delta kt) + \sinh^2(\pi\Delta k't))] \exp(-\pi t / \xi'_0) \qquad [VII.1] \, (VII.26)$$

in which ξ'_0 and ξ'_0 are the absorption distances and

$$\Delta k' = [1 + (s\xi_g)^2]^{1/2} \, \xi'_g{}^{-1} \qquad (VII.27)$$

The term $\exp(-\pi t/\xi'_0)$ represents the average absorption. It does not depend on s and describes the intensity attenuation caused by the specimen thickness.

The term $\sin^2\beta \, \sinh^2(\pi\Delta k't)$ is related to the anomalous absorption. It has a maximum value at $s = 0$ and cancels out for large s values.

As shown on figure VII.10b, the diffracted rocking curve is not much modified by absorption and remains perfectly symmetrical. Only the background is affected by the term $\sin^2\beta \, \sinh^2(\pi\Delta k't)$.

On the contrary, the transmitted rocking curve is strongly affected by absorption and becomes asymmetrical with respect to the value $s = 0$. This effect is quite visible on figure VII.10a.

VII.4 - Comparison of the kinematical and dynamical theories

The main differences between these two theories appear clearly on the simulations on figure VII.11. They show the kinematical and dynamical variations of the diffracted intensity with respect to s and t. The differences are especially visible for $s = 0$. In the case of the dynamical theory, the diffracted intensity cancels out each time the specimen thickness t is an integer number of extinction distances ξ_g.

In the kinematical theory, a diffracted intensity is observed whatever the specimen thickness. Note that the kinematical and dynamical behaviours are similar when s becomes large, i.e. far from the exact Bragg orientation.

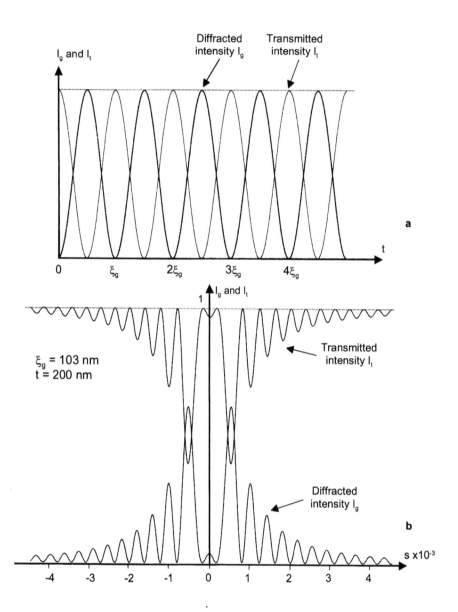

Figure VII.9 - Two-beam dynamical theory. The diffracted and transmitted intensities are complementary. Absorption is not taken into account.
a - The curves $I_g = f(t)$ (in bold line) and $I_t = f(t)$ (in fine line) are complementary.
b - The transmitted (bold line) and diffracted (fine line) rocking curves are complementary.

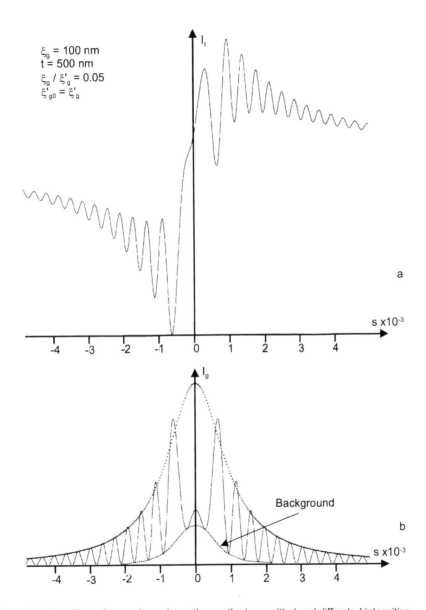

Figure VII.10 - Effect of anomalous absorption on the transmitted and diffracted intensities.
a - Transmitted rocking curve. The curve is strongly modified by absorption and is asymmetric.
b - Diffracted rocking curve. The curve remains symmetrical with respect to s = 0. The background intensity is modified around s = 0.

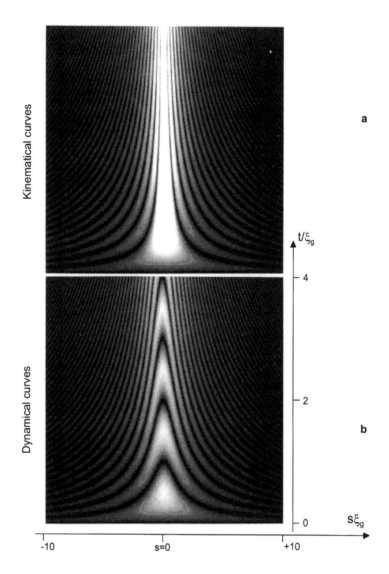

Figure VII.11 - Comparison between the kinematical and the dynamical theory under two-beam conditions. Curves $I_g = f(s\xi_g$ and $t/\xi_g)$.
a - Kinematical theory.
b - Dynamical theory. For s = 0, the diffracted intensity is zero when $t = k\xi_g$, where k is an integer.
Note that the dynamical and kinematical curves are similar when s is large.
Simulations. Courtesy of D. Bird.

Thus, the kinematical theory is only valid when:
- the specimen thickness t is very small,
- the extinction distance ξ_g is very large,
- the deviation parameter s is large.
We can consider that the kinematical theory is only valid when $t < \xi_g / 3$. In all the other cases, the dynamical theory should be used.

VII.5 - Appearance of the excess and deficiency lines

Careful examination of the CBED, Kossel and LACBED patterns shows that the deficiency and excess lines can be sorted into two main categories (Figure VII.12).

- The first category includes broad lines made up of fringes (Figures VII.12a, c and d). Usually, they are the brightest lines of the pattern and correspond to small hkl Miller indices and small extinction distances ξ_g. The fringe spacing depends on the specimen thickness. The excess lines are always symmetrical with respect to the line s = 0 but their intensity does not always have a maximum value at s = 0 (Figures VII.12a and d). As a result of anomalous absorption, the corresponding deficiency lines are more or less asymmetrical (Figures VII.12a and c).

- The second category includes sharp lines without fringes and of weak intensity. Generally, they have high hkl indices and large extinction distances (Figure VII.12b). They do not depend on the specimen thickness.

Thus, the appearance of the excess and deficiency lines depends mainly on two parameters:

- the extinction distance ξ_g. This parameter is related both to the nature of the (hkl) diffracting planes and to the beam wavelength λ,
- the specimen thickness t.

In order to understand the character of these lines, the two-beam kinematical and dynamical rocking curves described in the previous paragraph can be directly used, for the following reasons.

- In chapter IV and on figure VII.1b, we indicated that a map of the transmitted and diffracted intensities is observed in the transmitted and diffracted disks of a CBED pattern. In the diffracted disk, the deviation parameter s varies along lines parallel to the s = 0 line, the latter being perpendicular to the reciprocal vector $\mathbf{g}_{hkl}$. Along a disk diameter perpendicular to the excess line, the variation of the diffracted intensity as a function of s is directly observed, i.e. the rocking curve.

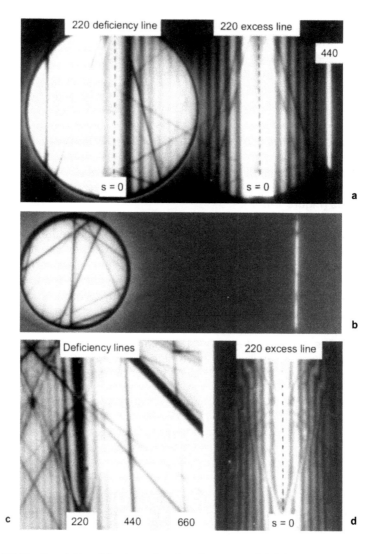

Figure VII.12 - Description of the two categories of excess and deficiency lines observed on a silicon specimen.
a - CBED pattern. The 220 lines are broad and made up of a set of fringes. The excess line is symmetrical with respect to s = 0. As a result of anomalous absorption, the deficiency line is asymmetrical.
b - CBED pattern. The excess and deficiency lines, with high hkl indices, are sharp.
c - Bright-field LACBED pattern. The 220 deficiency line consisting of fringes is asymmetrical. The 440 and 660 lines are sharp.
d - Dark-field LACBED pattern. The 220 excess line is symmetrical with respect to s = 0.

- As demonstrated in the next paragraph dealing with interactions, the two-beam conditions are satisfied along an excess line (except at the intersection with another line or in the close vicinity of another line).

Thus a direct correlation can be established between the diffracted intensity inside a disk and the rocking curve under two-beam conditions (Figure VII.13).

VII.5.1 - Effect of the extinction distance ξ_g on the excess and deficiency lines

The examples of the 400, 800 and 16 0 0 excess lines observed under two-beam conditions in a silicon specimen of approximate thickness 200 nm are given on figures VII.13a and VII.14a and b. These lines have quite different extinction distances ξ_g (170, 463 and 1470 nm respectively at 300 kV). The 400 line, and to a lesser extent, the 800 line are made up of many fringes. The 16 0 0 line is sharper and displays only a single maximum. The corresponding rocking curves, calculated by means of the kinematical and dynamical theories are also shown on these figures.

The dynamical theory gives results in perfect agreement with the experimental patterns whatever the line is. In particular, it reproduces perfectly the 400 line. The kinematical theory is valid only for the 800 and 16 0 0 reflections, i.e. for reflections having a large extinction distance.

VII.5.2 - Effect of the specimen thickness on the excess and deficiency lines

VII.5.2.1 - CBED and Kossel patterns

For CBED and Kossel patterns, the probe size S is very small so that the specimen thickness can be considered to be constant in the diffracted area. For the lines made up of fringes, the number of fringes as well as the fringe spacings are thickness-sensitive.

This effect is clearly highlighted on the dark field CBED patterns on figure VII.15. These patterns display the 400 excess line from a silicon specimen for the two thickness values t = 101 and 191 nm. The corresponding dynamical rocking curves are in perfect agreement.

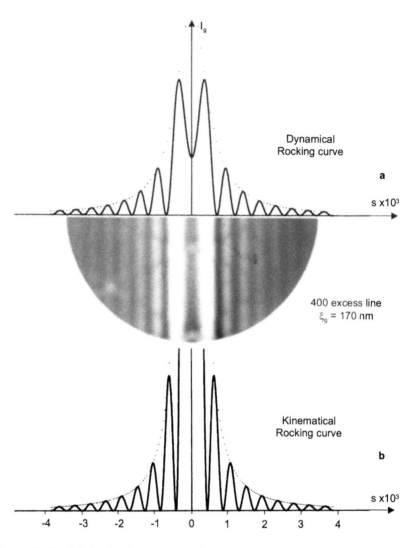

Figure VII.13 - Relationship between the diffracted intensity in a CBED diffracted disk and the rocking curve under two-beam conditions. Case of the 400 line from a silicon specimen with an extinction distance $\xi_g = 170$ nm at 300 kV.
a - Dark-field CBED pattern and corresponding dynamical rocking curve. The agreement is excellent.
b - Kinematical rocking curve. The agreement with the experimental pattern is poor especially in the central area of the pattern. It only becomes correct for large values of s.

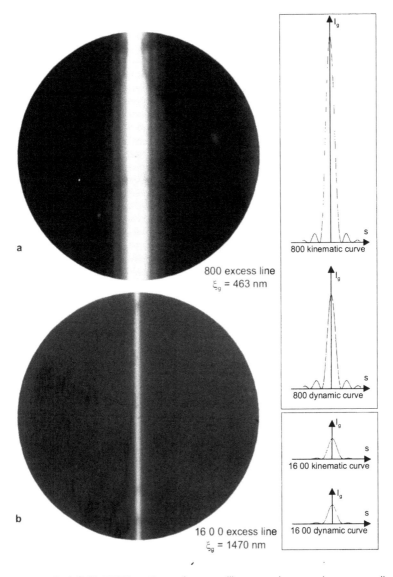

Figure VII.14 - Dark-field CBED patterns from a silicon specimen and corresponding kinematical and dynamical rocking curves.
a - 800 excess line with an extinction distance ξ_g = 463 nm (at 300 kV).
b - 16 0 0 excess line with an extinction distance ξ_g = 1470 nm (at 300 kV).
For these excess lines with large extinction distances, both the dynamical and kinematical rocking curves are in good agreement with the experimental patterns.

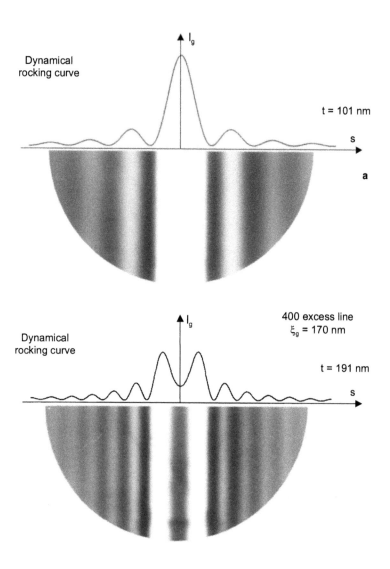

Figure VII.15 - Effect of the specimen thickness t on the "dynamical" lines. 400 dark-field CBED patterns and corresponding dynamical rocking curves observed in a silicon specimen.
a - Specimen thickness: t = 101 nm.
b - Specimen thickness: t = 191 nm.
The agreement with the dynamical rocking curves is excellent.

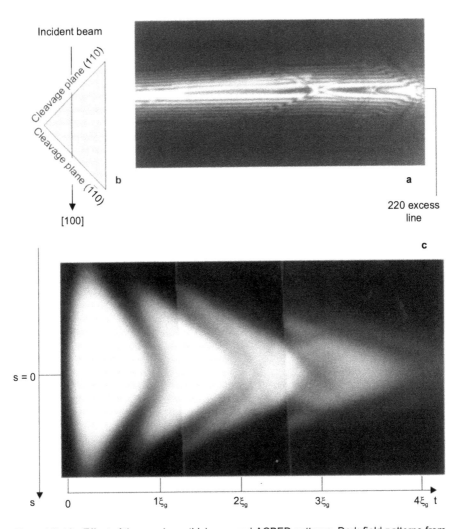

Figure VII.16 - Effect of the specimen thickness on LACBED patterns. Dark-field patterns from a silicon specimen.
a - 220 excess lines. The specimen displays thickness variations, which result in the appearance of a new fringe each time the specimen thickness t increases by an extinction distance.
b - Cleaved wedge-shaped specimen. The thickness increases linearly.
c - 400 excess line in a wedge-shaped specimen. The appearance of the line reproduces the dynamic simulation on figure VII.11b.
Silicon specimen. Courtesy of D. Laub.

VII.5.2.2 - LACBED patterns

Since these patterns are obtained in a defocus mode, a larger specimen area is observed. If this area is perfect, i.e. if its thickness and its orientation are constant, then the effect of the specimen thickness is identical with that for CBED and Kossel patterns. If the illuminated area contains thickness variations, the appearance of the fringes is disturbed. New fringes appear each time the specimen thickness is increased by one extinction distance. This effect is visible on figure VII.16a.

Cleaved wedge-shaped specimens form a particularly interesting case. In these specimens, the thickness varies linearly as a function of the distance from the edge of the specimen (Figure VII.16b). The Bragg lines observed on these specimens (Figure VII.16c) reproduce the simulations shown on figure VII.11.

VII.5.3 - Angular width of the excess and deficiency lines

The angular width $\Delta\theta$ of the excess and deficiency lines can be estimated by means of figure VII.17, which indicates that:

$$\Delta\theta = 2/g_{hkl} / \xi_g \qquad\qquad (VII.28)$$

It is interesting to compare this value with the convergence angle and the Bragg angle by means of the ratios $\Delta\theta/2\alpha$ and $\Delta\theta/\theta_B$.

Application

The calculation of $\Delta\theta$ for the 400 silicon line (with interplanar spacing $d_{400} = 0.135$ nm and extinction distance $\xi_{400} = 170$ nm) obtained at 300 kV for an incident electron beam of convergence $2\alpha = 0.5°$ gives:

$\Delta\theta = 2\times0.135/170 = 1.5\,mrad$

$2\alpha = 0.5° = 9\,mrad$

$\theta_{400} = 0.41° = 7\,mrad$

$\Delta\theta/2\alpha = 0.18$

$\Delta\theta/\theta_{400} = 0.15$

This means that the 400 excess line occupies an appreciable fraction of the diffracted disc (18% of the diameter) (Figure VII.12a). $\Delta\theta$ is about 6 times smaller than the Bragg angle θ_{400}.

For the 800 silicon line (with spacing $d_{800} = 0.068$ nm and $\xi_{800} = 463$ nm) we have:

$\Delta\theta = 2\times0.068/463 = 0.3\,mrad$

$2\alpha = 0.5° = 9\,mrad$

$\theta_{800} = 0.830° = 14\,mrad$

$\Delta\theta/2\alpha = 0.03$

$\Delta\theta/\theta_{800} = 0.02$

This time, the 800 line is much narrower than the disk diameter (3% of the disk diameter) (Figure VII.13a). $\Delta\theta$ is about 50 times smaller than the Bragg angle θ_{800}.

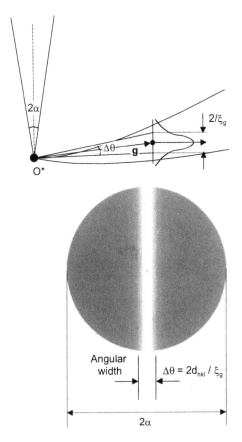

Figure VII.17 - Determination of the angular width $\Delta\theta$ of the excess line in the diffracted disk.

VII.5.4 - Dynamical and quasi-kinematical lines

Taking into account the previous considerations, we can sort the excess and deficiency lines into two categories.

- **Sharp lines**. These display a single main maximum (or a minimum for the deficiency lines) for s = 0. These lines, which can be simulated using the two-beam kinematical theory, will be called **"quasi-kinematical" lines**. Since they have a kinematical or quasi-kinematical behaviour, the diffracted intensity, for s = 0, is related to the squared modulus of the structure factor $|F_g|^2$. These lines can be used to carry out structure determinations. When these lines correspond to a very large extinction distance, they become very sharp and can be used for accurate measurement of the lattice parameters or for accurate identification of the crystal orientation. We give several examples in chapter XI.

- **Broad lines**. These display a more or less significant number of fringes and can be simulated only by using the dynamical theory. They will be called **"dynamical lines"**. Since the interfringe spacing greatly depends on the thickness, they can be used to measure the specimen thickness. When the specimen is very thin, the interfringe spacing becomes very large and the diffracted intensity is nearly constant within the disk. No details are visible inside the disks, as shown on the example on figure VII.18.

At a first approximation, we can consider that the transition between the dynamical and the kinematical behaviour occurs when the following condition is satisfied:

$$t < \xi_g /3.$$

Note
The classification of the lines into dynamical and quasi-kinematical lines is rather artificial since all the lines can be interpreted with the dynamical theory and furthermore true kinematical behaviour occurs only in rare circumstances. However, we continue to use this classification, because it has the advantage of simplifying the pattern indexing.

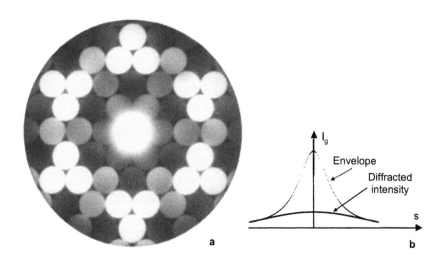

Figure VII.18 - CBED pattern from a Ru_7B_3 specimen with low thickness. The diffracted intensity is nearly constant inside each disk and no details are visible.
a - Experimental [001] zone-axis pattern.
b - Rocking curve in the case of small specimen thickness and a small extinction distance. The diffracted intensity varies very little with respect to s.

VII.6 - Dynamical Interactions

Dynamical interactions play an important role in convergent-beam electron diffraction and produce typical effects. First, we examine the dynamical interactions under two-beam conditions.

VII.6.1 - Interactions under two-beam conditions

VII.6.1.1 - Parallel incident beam

Figure VII.19 describes the formation of a diffraction pattern under two-beam conditions in the case of a parallel incident beam. This incident beam produces a transmitted beam and a single diffracted beam. In the back focal plane of the objective lens, the diffraction pattern consists of two spots A_T and A_D separated by a distance $D = 2f\theta$. Interactions can occur only between the diffracted beam and the single transmitted beam. On the following diagrams, we represent the absence of interaction (kinematical behaviour) by a single arrow and the presence of interactions by a double arrow (dynamical behaviour).

VII.6.1.2 - Convergent Incident beam

When the incident beam is convergent, each incident ray behaves like the single parallel beam described in the previous paragraph. Thus, each of the two incident rays A_E and B_E drawn on figure VII.20, gives a pair of transmitted and diffracted points separated by a distance $D = 2f\theta_B$. These are the A_T and A_D pair for the AE incident ray, and the B_T and B_D pair for the BE incident ray. The two points of a pair are located at equivalent places in the transmitted and diffracted disks, since they are separated by the same distance D as the centres of these disks. This property has a major consequence for the interactions.

Interactions will occur only between equivalent points, for example, between the two points A_T and A_D, or between B_T and B_D, as shown on figure VII.20.

On the other hand, interaction are impossible between two non-equivalent points because these points come from different incident rays; this is the case for the points A_T and B_D or for the points A_D and B_T. They cannot interact. From this point of view, the convergent beam diffraction patterns can be regarded as being made up of a collection of spot patterns (this approach was already developed in the paragraph IV.1.1.1 and shown on figure IV.3).

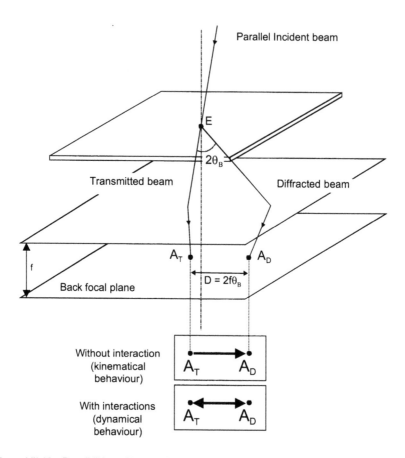

Figure VII.19 - Possibilities of interactions in the case of a diffraction pattern produced by a parallel incident beam under two-beam conditions. Interactions can occur only between the transmitted beam and the single diffracted beam. Dynamical interactions are schematised by a double arrow and absence of interaction (kinematical behaviour) by a single arrow.

VII.6.1.3 - Large-angle convergent beam: Kossel patterns

What happens in the case of the large convergence angle used for the Kossel patterns? The transmitted and diffracted disks overlap partially as shown on figure VII.21. The points located outside the superimposition area have the same possibilities of interactions as those described previously for a convergent beam. This is the case for the points C_T and C_D on figure VII.21a.

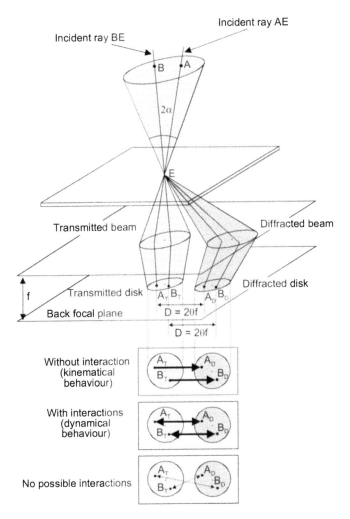

Figure VII.20 - Possibilities of interactions in the case of a CBED pattern. Dynamical interactions can occur only between equivalent points inside the disks (for example, A_T and A_D or B_T and B_D) because these points come from the same incident ray. Interactions between non-equivalent points (for example, A_T and B_D or A_T and A_D) are impossible because these points come from different incident rays.

The points located inside the overlapping area have a more complicated behaviour. We consider the point P on figure VII.21b. In fact, this point is made up of the superimposition of:

- a transmitted point A_T, which can interact with its equivalent diffracted point A_D,
- a diffracted point B_D, which can interact with its equivalent transmitted point B_T.

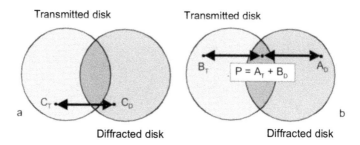

Transmitted disk   Transmitted disk

a

b

Diffracted disk   Diffracted disk

Figure VII.21 - Possibilities of interactions in the case of a large convergent beam under two-beam conditions. Kossel pattern. The two disks overlap partially.
a - All the equivalent points located outside the overlapping area interact like those on figure VII.20. This is the case for the points C_T and C_D.
b - Any point P located inside the overlapping area consists of the superimposition of a transmitted point A_T and a diffracted point B_D coming from different incident rays. If the incident beam is incoherent, the intensity of the point P is the sum of the intensities of the points A_T and B_D. If the beam is coherent, then the amplitudes are added.

Can these points A_T and B_D interact? They come from two different incident rays. If these two rays are **incoherent**, there is no possibility of interaction and the intensity obtained at the point P will be the sum of the intensities of the two points A_T and B_D. This behaviour is observed with electrons coming from tungsten or LaB_6 (lanthanum hexaboride) electron guns. On the contrary, if these rays are coherent, they can interact and then it is not the intensities that are added but the amplitudes. This is what happens with electrons from a field-emission gun. This type of pattern is encountered in **coherent convergent beam electron diffraction** [VII.2,3].

VII.6.2 - Many-beam interactions

VII.6.2.1 - Three-beam patterns

Three-beam patterns are the simplest cases of many-beam patterns. They consist of a transmitted disk and two diffracted disks. Several configurations of these three disks are possible. Two of them are shown on figure VII.22.

When the three disks do not overlap (Figure VII.22a), the interactions are the same as those described previously. They can occur only between **triplets of equivalent points** (for example, A_T, A_{D1} and A_{D2}) located in the three disks since each triplet comes from the same incident ray.

A new type of interaction is now observed; it concerns the two diffracted beams (for example the points A_{D1} and A_{D2} on figure VII.22a). Consider the case of a Kossel pattern for which the two diffracted disks overlap partially (Figure VII.22b). Any point P located in the overlapping zone corresponds to the superimposition of two diffracted points A_{D2} and B_{D1}. These two points interacting with their respective equivalent points come from two different incident rays. They can interfere only if the latter are coherent. If not, the intensity of the point P is simply the sum of the intensities of the diffracted points A_{D2} and B_{D1}.

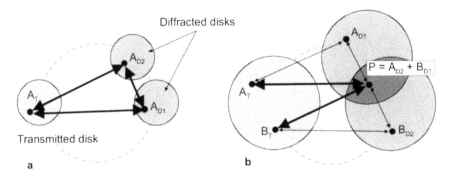

Figure VII.22 - Possibilities of interactions in the case of CBED and Kossel patterns under three-beam conditions.
a - CBED pattern. The transmitted disk and the two diffracted disks do not overlap. Interactions can occur between any triplets of equivalent points (for example, points A_{T1}, A_{D1} and A_{D2}) because they come from the same incident ray.
b - Kossel pattern. The two diffracted disks overlap partially. Any point P located in the overlapping area consists of the superimposition of two diffracted points A_{D2} and B_{D1} coming from two different incident rays. If the incident beam is incoherent, they cannot interfere and the intensity of the point P is the sum of the intensities of the two points A_{D2} and B_{D1}. If the incident beam is coherent, the amplitudes are added.

VII.6.2.2 - Multi-beam patterns

All the previous results apply to multi-beam patterns and to zone-axis patterns, which are a special case of multi-beam patterns (Figure VII.23).

For these zone-axis patterns, we distinguish.

- Interactions only involving the transmitted disk and the disks located in the zero order Laue zone. These interactions are called **two-dimensional interactions** because they only involve the zeroth layer of the reciprocal lattice. They produce the typical fringes observed in the central area on zone axis patterns. The pattern on figure VII.24a illustrates this case. The symmetry of these patterns is related to the ten "projection" diffraction groups (see reference i.3).

- Interactions concerning the reflections located in the higher-order Laue zones and the transmitted beam. These interactions are called **three-dimensional interactions** because they involve several layers of the reciprocal lattice. They give the sharp HOLZ lines observed in the transmitted disk of CBED patterns (Figure 24b) and are related to the 31 diffraction groups.

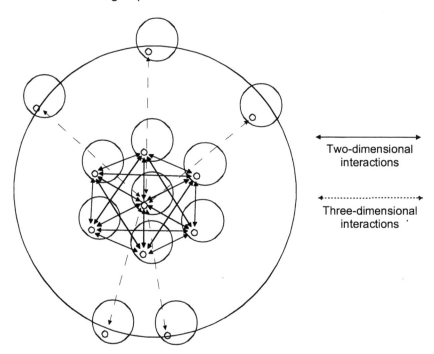

Two-dimensional interactions

Three-dimensional interactions

Figure VII.23 - Possibilities of interactions for a zone-axis CBED pattern. Two-dimensional interactions (continuous lines) occur between equivalent points located inside the disks of the zero-order Laue zone and the transmitted disk. Three-dimensional interactions (dotted lines) involve the high-order Laue zones and the transmitted disk. For the sake of clarity, the other interactions are not drawn.

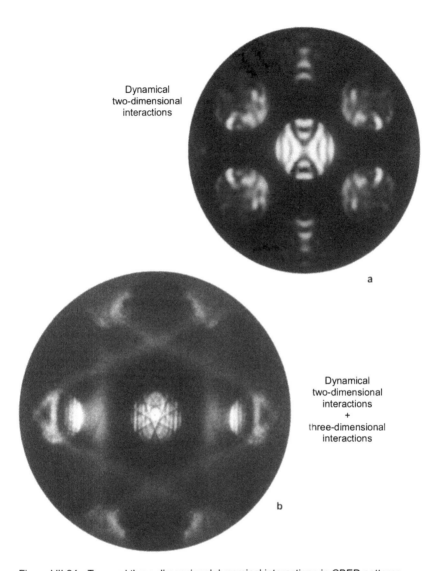

Dynamical
two-dimensional
interactions

a

Dynamical
two-dimensional
interactions
+
three-dimensional
interactions

b

Figure VII.24 - Two and three-dimensional dynamical interactions in CBED patterns.
a - <110> zone axis pattern from a silicon specimen. The pattern contains only zero-order Laue zone reflections. These reflections give two-dimensional dynamical interactions appearing in the form of a complex set of black and white broad fringes.
CBED pattern. Courtesy of A. Redjaïmia.
b - Central area of a <210> zone axis pattern from a ferrite specimen. The pattern contains, in addition to the two-dimensional dynamical interactions, sharp HOLZ lines which are produced by three-dimensional dynamical interactions coming from the high order Laue zones. Note the presence of Kikuchi lines outside the disks.

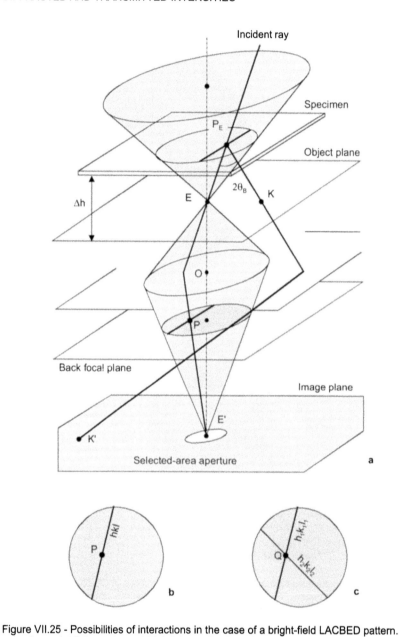

Figure VII.25 - Possibilities of interactions in the case of a bright-field LACBED pattern.
a, b - Any point P located on a hkl deficiency line is under two-beam conditions.
c - Any point P located at the intersection of two $h_1k_1l_1$ and $h_2k_2l_2$ deficiency lines consists of the superimposition of two diffracted points coming from the same incident ray. These two points can interfere and are under three-beam conditions.

VII.6.2.3 - LACBED patterns

The results given above remain valid for LACBED patterns where the transmitted or a diffracted beam is selected with the selected-area aperture. The selection of the beams takes place after the possible interactions have already occurred in the specimen.

For example, we consider a point P located on a deficiency line of a LACBED pattern (Figure 25b). This point P comes from an incident ray crossing the specimen at the point P_E. At this point, **a transmitted beam and a single diffracted beam are produced; this means that the two-beam conditions are locally satisfied**.

Now, we consider a point Q located at the intersection of two Bragg lines $h_1k_1l_1$ and $h_2k_2l_2$ (Figure VII.25c). It comes from an incident ray giving a transmitted ray and two diffracted rays $h_1k_1l_1$ and $h_2k_2l_2$. This is a three-beam configuration. The transmitted beam and the two diffracted beams can interact by giving quite specific aspects.

Some examples are given on figures VII.26a, b and c. As expected, these effects only occur at the intersections of "dynamical" lines. The intersections of quasi-kinematical lines are not disturbed (Figure VII.26d) because the intensity at these intersections is simply the sum of the intensities of the two involved lines.

These three-beam configurations are particularly useful in **quantitative electron diffraction**. They allow, in an elegant way, the determination of the structure factor. The procedure is described in detail in the book "Electron Microdiffraction" of J.C.H Spence and J.M. Zuo [VII.4] and in some specialized papers [VII.5, 6].

A particularly interesting three-beam configuration occurs in connection with **"kinematically forbidden" lines**, i.e. Bragg lines with a zero structure factor F_{hkl} arising from the presence of screw axes and/or glide planes in the crystal structure. In conventional electron diffraction (SAED), kinematically forbidden reflections appear frequently on diffraction patterns. They result from dynamical interactions and are hence not easily identified among the other allowed reflections. On the contrary, they can be easily identified from the Gjønnes and Moodie lines [i.4] observed on CBED patterns when the crystal has specific orientations with respect to the incident beam.

On LACBED patterns, kinematically forbidden lines only appear at the intersections with "dynamical" lines where multiple diffraction paths occur [VII.7-9].

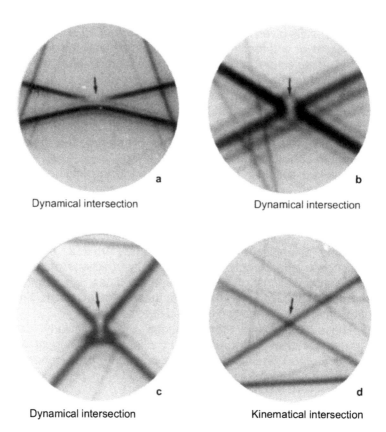

a

Dynamical intersection

b

Dynamical intersection

c

Dynamical intersection

d

Kinematical intersection

Figure VII.26 - Detailed views of some intersections of two Bragg lines present on bright-field LACBED patterns.

a, b, c - Intersections of two dynamical lines. They interfere very strongly and produce typical effects.

d - Intersection of two quasi-kinematical lines. They do not interfere and the intensity at the intersection is simply the sum of the intensities of the two Bragg lines.

The example of the 420 line from a silicon specimen is given on figure VII.27a. This line is kinematically forbidden because of the presence of diamond glide planes in the silicon structure. It only appears at the intersections with some allowed lines where dynamical interactions occur. Everywhere else it disappears. This effect is even more visible on dark-field LACBED patterns (Figure 27b) where the forbidden lines look like dotted lines since they only appear at the intersections with allowed reflections.

This property can be used to detect the kinematically forbidden reflections very reliably and then to identify glide planes and screw axes. The latter give access to the space group.

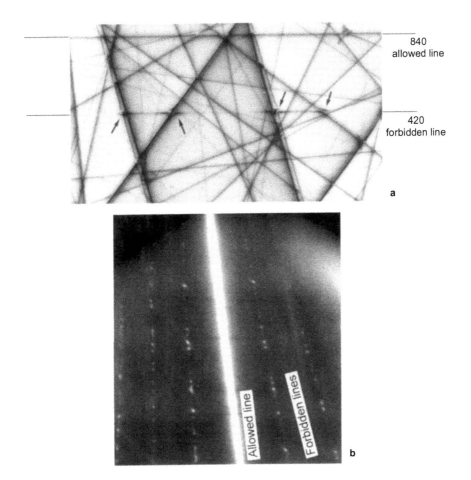

840
allowed line

420
forbidden line

a

b

Figure VII.27 - Kinematically forbidden lines.
a - The 240 line is kinematically forbidden because of the presence of diamond glide planes in the crystal structure. It only appears at some intersections (marked with an arrow) with dynamical Bragg lines where dynamical interactions occur. Apart from these intersections, the line is invisible. Note that the allowed line 840 is continuous and has a normal behaviour.
b - Dark-field LACBED pattern. Some forbidden lines look like dotted lines since they only appear at the intersections with allowed lines.

CHAPTER VIII - Experimental methods

Kossel and LACBED techniques require complex microscope alignment. The very large convergences used to obtain these patterns produce strong spherical aberration and lead to unusual features for microscopists more accustomed to work with parallel incident beams. The systematic use of the nanoprobe mode can also be a difficulty since this mode generally operates with parallel incident beams and not with widely convergent beams.

It is therefore useful to give a description of the usual operating modes of the electron microscope before describing the specific operating conditions involved in the LACBED technique.

VIII.1 - Operating principle of magnetic lenses

In chapter II, we saw that the magnetic lenses of the microscope can be regarded as thin convergent optical lenses. However, unlike glass lenses, their focal length can be varied continuously by changing their excitation current. With a high excitation current, electrons are more strongly deviated and the focal length decreases.

Diagrams a, b, c and d on figure VIII.1 show that a change of the focal length f makes it possible to magnify or demagnify the image A'B' of an object AB located at the distance u from the lens. The image is magnified if $u > f > u/2$ and demagnified if $f < u/2$.

We note that a change of the focal length also causes a change of the position of the image plane. This plane is located at the distance v from the lens with $v = \gamma u$, γ being the magnification of the lens.

This description concerns magnetic lenses designed to form the image of an object located at a given distance u from the lens (they will be called type I lenses). An example is the C_1 condenser, whose role is to provide a demagnified image of the electron gun. The latter is separated from the C_1 condenser by a given distance u.

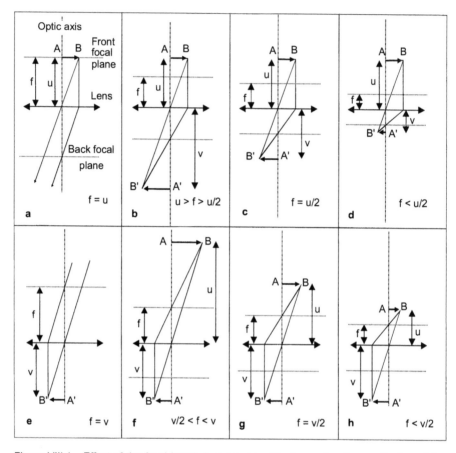

Figure VIII.1 - Effect of the focal length f on the magnification or the demagnification of the image A'B' of an object AB.

a, b, c, d - Type I lens. The object AB is located at a given distance u from the lens. Its image A'B', magnified or demagnified, is formed at a variable distance v from the lens.

e, f, g, h - Type II lens. The image A'B', magnified or demagnified, is located at a given distance v from the lens. The object AB is located at a variable distance u from the lens.

A second type of lenses (called type II lenses) consists of magnetic lenses designed to provide an image A'B' located at a given **distance v** from the lens. For example, the final image (or the final diffraction pattern) formed by the projector lens must be in the plane of the screen located at a given distance v from this lens. For this second type of lens, the position u of the object AB depends on the focal length f (Figures VIII.1e, f, g and h). The image is magnified if f < v/2 and demagnified if v/2 < f < v.

VIII.1.1 - Focused, under-focused and over-focused lenses

A lens is focused when the image of an object is formed in a given image plane (Figure VIII.2b). If the image is formed below or above this plane, the lens is **under- or over-focused** respectively (Figures VIII.2a and c).

An accurate focus is obtained by modifying the focal length f, i.e. the excitation current of the magnetic lens. A low excitation current corresponds to a large focal length and to an under-focused lens (Figure VIII.2a). Conversely, a high current corresponds to a small focal length and to an over-focused lens (Figure VIII.2c).

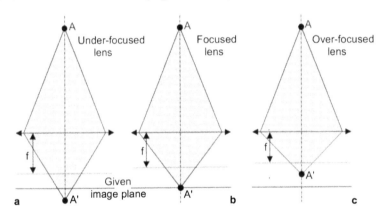

Figure VIII.2 - Description of under-focused, focused and over-focused lenses.
a - Under-focused lens. The image A' of a point A is formed below the given image plane. The focal length f is large and the excitation current small.
b - Focused lens. The image A' is exactly formed in the given image plane.
c - Over-focused lens. The image A' is formed above the given image plane. The focal length f is small and the excitation current is high.

VIII.1.2 - Condenser-objective lens

Throughout this chapter we assume that the objective lens of the microscope is a **condenser-objective lens** of the Rieke-Ruska type. This type of magnetic lens equips most of the recent analytical microscopes and gives very large convergence angles as well as very small probe sizes. This lens may be regarded as two identical magnetic lenses symmetrically disposed on either side of a symmetry plane (Figure VIII.3a). The specimen is located in this symmetry plane, which hence constitutes the object plane of the objective lens. Moreover, this plane also corresponds to the **eucentric height***.

Modern microscopes are equipped with a eucentric specimen stage. With such a stage, the specimen can be tilted while keeping the illuminated specimen area on the optic axis of the microscope.

The magnetic lens located above the specimen acts as a third condenser lens and will be called the C_3 condenser. The lower magnetic lens constitutes the true objective lens. This type of condenser-objective lens is used in the "nanoprobe" mode of the Philips-FEI electron microscopes and is called a "twin" lens. We shall often use this term twin, since most of the experimental patterns reproduced in this book were obtained with the twin lens of a Philips CM30 transmission electron microscope.

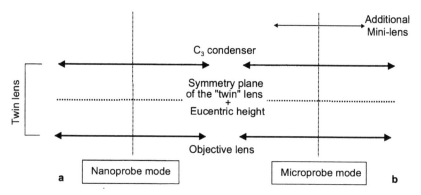

Figure VIII.3 - Schematic description of a twin condenser-objective magnetic lens. The upper part of the twin lens constitutes a third condenser lens (C_3 condenser) and the lower one, the objective lens itself. The specimen is located in the symmetry plane of the twin lens, which constitutes the object plane of the objective lens. It also corresponds to the eucentric height of the specimen stage.
a - Twin lens in the nanoprobe mode.
b - Twin lens in the microprobe mode. An additional mini-lens is brought into operation and allows larger specimen areas to be observed.

VIII.1.3 - Description of the "nanoprobe" and "microprobe" modes

Two different operating modes are available on Philips-FEI microscopes: the "nanoprobe" and the "microprobe" modes. The nanoprobe mode produces incident beams having very large convergence angles as well as very small probe sizes. This mode is systematically used in the LACBED technique. It has a serious disadvantage: the maximum size of the illuminated area of the specimen is about one micron. It cannot be used at low magnifications, therefore.

In order to remove this limitation, an additional mini-lens located above the C_3 condenser operates in the microprobe mode, which is the normal mode of the microscope (Figure VIII.3b). This mini-lens increases the size of the illuminated area but, on the other hand, it decreases the convergence angle and increases the probe size. The microprobe mode cannot be used to perform LACBED experiments.

VIII.2 - General description of the microscope

In the following section, we assume that the microscope is operating in the nanoprobe mode and that the specimen is located at its normal position in the microscope, i.e. in the object plane of the objective lens (at the eucentric height).

From the optical point of view, electron microscope imaging can be divided into four stages:
- the specimen illumination, which is controlled by the condenser system composed of three magnetic lenses in the case of a twin lens,
- the formation of the first image and diffraction pattern, which is performed by the objective lens,
- the magnification either of the image or of the diffraction pattern, which is modified by the intermediate lens,
- the projection of the image or of the diffraction pattern on the screen, which is carried out by the projector lens.

The main elements belonging to each of these four stages are shown on figure VIII.4.

Note
The arrangement and the number of magnetic lenses depend on the type of microscope. The number of condenser, intermediate and projector lenses can be greater than the one given on figure VIII.4.

Some elements are located at well-defined positions inside the microscope column. They are:
- the electron gun that provides the electrons. Here we consider the **crossover*** to be the electron source.
- the magnetic lenses,
- the four main apertures:
    - the C_1 aperture,
    - the C_2 aperture,

*The electrons emitted by the gun are concentrated by an electrode called Wehnelt; they then pass through a disk of least diameter called the crossover.

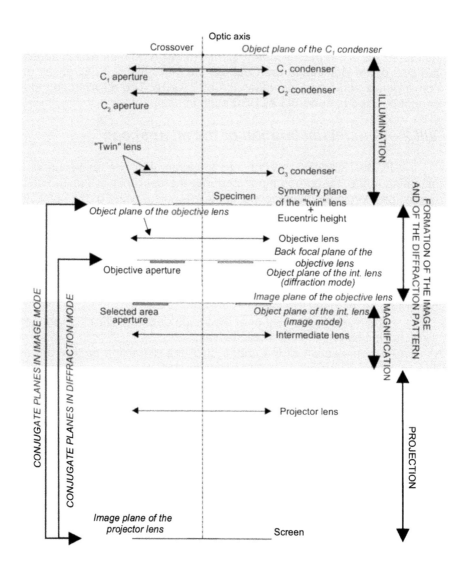

Figure VIII.4 - Description of transmission electron microscope imaging into four stages:
- the specimen illumination,
- the formation of the image and diffraction pattern,
- the magnification either of the image or of the diffraction pattern,
- the projection either of the image or of the diffraction pattern on the screen.

- the objective aperture located in the back focal plane of the objective lens,
- the selected-area aperture located in the image plane of the objective lens,
- the screen,
- the specimen located at the eucentric height. This height corresponds to the symmetry plane of the twin lens.

Consequently:
- the position of the object plane of the C_1 condenser is given and must be in the crossover plane,
- the position of the object plane of the objective lens is given and must be at the eucentric height,
- the position of the back focal plane of the objective lens is given and corresponds to the position of the objective aperture,
- the position of the image plane of the objective lens is given and corresponds to the position of the selected-area aperture,
- the position of the object plane of the intermediate lens is given. In the image mode, it is located in the plane of the selected-area aperture (the image plane of the objective lens). In the diffraction mode, it is located in the plane of the objective aperture (the back focal plane of the objective lens).
- the position of the image plane of the projector lens is given and corresponds to the screen plane.

Note
We assume here that the focal length of the objective lens is given and adjusted so that its object plane is located at the eucentric height and its image plane corresponds to the plane containing the selected-area aperture. In the image mode, this assumption implies that specimen focusing is carried out by adjusting the height of the specimen and not by adjusting the objective lens current. This is the solution adopted on recent microscopes, where the objective lens operates at optimal conditions at a given excitation current.

In the image mode, the specimen and the screen are conjugate planes.

In the diffraction mode, the back focal plane of the objective lens and the screen are conjugate planes.

Note
*Many microscopes are equipped with a **diffraction lens** located between the objective lens and the intermediate lens. This allows accurate focusing of the diffraction pattern by making the back focal plane of the objective lens exactly conjugate to the object plane of the intermediate lens (and the screen). For the sake of clarity, this lens is not shown on the figures in this book.*

VIII.2.1 - Description of the condenser

Usually, the condenser consists of two magnetic lenses called C_1 and C_2. We recall that, in the case of a twin lens, the upper part of the twin lens acts as a third condenser lens C_3. This type of condenser will be described here.

The aim of the condenser is to illuminate the specimen. With two or three lenses, it is possible to obtain parallel, convergent or divergent incident beams (Figures, VIII.5a, b and c). Moreover, it is also possible to adjust independently the probe size S and the convergence semi-angle α of the beam (Figures VIII.5d, e, f and g).

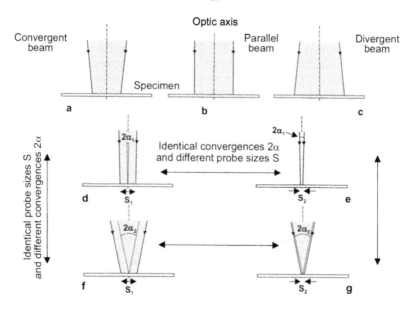

Figure VIII.5 - Schematic descriptions of the illumination conditions produced by the condenser.
a, b, c - Convergent, parallel and divergent incident beams.
d, e, f, g - Convergent incident beams with convergence semi-angles α_1 and α_2 and with probe sizes S_1 and S_2. The probe size S and the convergence semi-angle α can be adjusted independently using the C_1 and C_2 condensers.

VIII.2.1.1 - Effect of the C_1 condenser

The first condenser C_1 is used to modify the probe size S. By increasing the excitation current, its focal length f_{C1} decreases and a strongly demagnified image A' of the crossover A is formed (Figure

VIII.6). This is a type I lens (lens with a given u value) and its focal length is $f_{C1} < u/2$ (Figure VIII.1d).

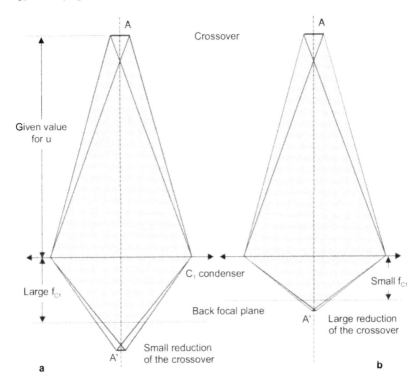

Figure VIII.6 - Effect of the C_1 condenser on the demagnification of the image A' of the crossover A.
a - A low excitation current gives a large focal length f_{C1} and a weak demagnification of the crossover image A'.
b - A high excitation current gives a small focal length f_{C1} and a strong demagnification of the crossover image A'.

VIII.2.1.2 - Effect of the C_2 condenser

The C_2 condenser is used to illuminate the specimen with a divergent, parallel or convergent incident beam. It forms a second crossover image A'' whose position depends on its focal length f_{C2} and therefore on its excitation current. The condenser C_3 then forms a third crossover image A'''.

In order simplify the following drawings, we assume that the crossover is a point.

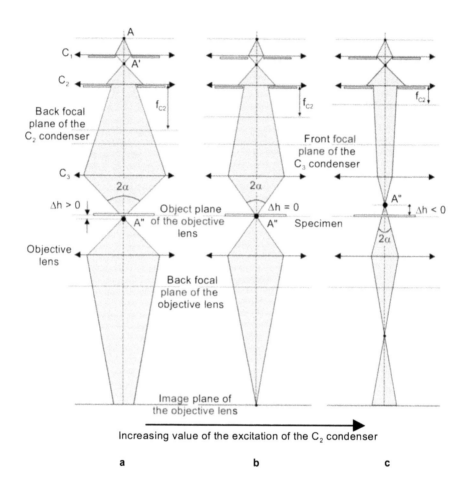

Increasing value of the excitation of the C_2 condenser

a   b   c

Figure VIII.7 - Evolution of the illumination conditions of the specimen when the excitation of the C_2 condenser increases (and hence the focal length f_{c2} decreases).
a - Very low excitation current. The crossover image A" is formed below the specimen ($\Delta h > 0$). The incident beam is convergent and under-focused.
b - Low excitation current. The crossover image A" is formed exactly in the object plane of the objective lens ($\Delta h = 0$). The incident beam is convergent and perfectly focused. This is the configuration used to perform CBED and Kossel experiments.
c - Medium excitation current. The crossover image A" is formed above the object plane of the objective lens ($\Delta h < 0$). The incident beam is divergent and over-focused.

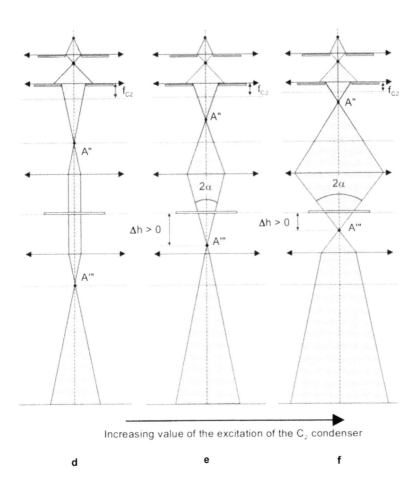

Increasing value of the excitation of the C$_2$ condenser

d   e   f

Figure VIII.7 - Continuation.
d - Medium excitation current. The crossover image A" is formed exactly in the front focal plane of the C$_3$ condenser lens. The specimen is illuminated by a parallel incident beam. This is the configuration used in conventional imaging and diffraction.
e, f - High and very high excitation currents. The crossover image A" is formed above the front focal plane of the C$_3$ condenser lens. The incident beam is convergent and under-focused. The third crossover image A''' is located below the specimen ($\Delta h > 0$).

195

Figures VIII.7a to f, describe how the illumination conditions change as the excitation of the C_2 condenser increases. The focal length f_{C2} decreases and the incident beam changes continuously, from a convergent beam to a divergent beam, then to a parallel beam and finally to a convergent beam.

If the crossover image A" is formed exactly in the object plane of the objective lens, then the condenser is perfectly focused and a large convergent beam is obtained in this plane (Figure VIII.7b). **This particular configuration is used mainly to generate Kossel and LACBED patterns.**

If the crossover image A" is formed exactly in the front focal plane of the C_3 condenser, then the specimen is illuminated by a parallel incident beam (Figure VIII.7d). **This configuration is used for conventional imaging and diffraction**.

In all other cases, the condenser is under or over-focused (Figures VIII.7a, c, e and f) and both the convergence semi-angle α and the position of the crossover images A" or A''' are modified. They are located above ($\Delta h < 0$) (Figure VIII.7c) or below ($\Delta h > 0$) (Figures VIII.7a, e and f) the object plane of the objective lens. **These configurations are used to create "eucentric" LACBED, CBIM and "defocus" CBED patterns**.

Figures VIII.8a to f show how the image of the illuminated area evolves, in the absence of specimen, as the excitation of the C_2 condenser is increased. The diameter of this area decreases, passes through a minimum when the condenser lens is perfectly focused (Figure VIII.8b) and then increases again.

VIII.2.1.2.1 - Effect of the C_2 aperture

Effect of the size of the C_2 aperture

The size of the C_2 aperture is used to change the convergence semi-angle α as shown on figure VIII.9. Its effect is very important in CBED. In LACBED, the largest available C_2 aperture is used.

Note
On some electron microscopes, the convergence semi-angle α can also be selected directly by means of the condenser lenses (α selector). This solution allows a continuous change of the value of the convergence angle.

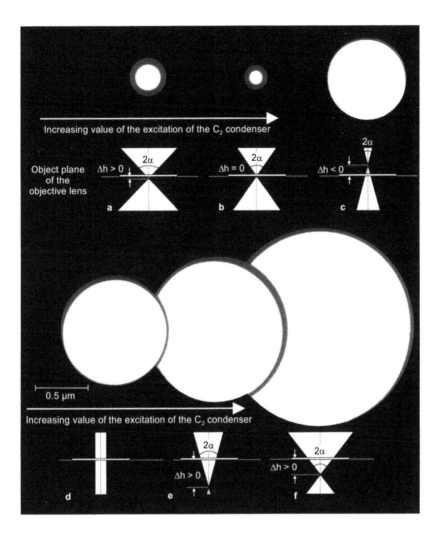

Figure VIII.8 - Evolution of the image of the illuminated area as the excitation of the C_2 condenser increases. These images, obtained without any specimen, correspond to the diagrams on figure VIII.7. The size of the illuminated area passes through a minimum (at b) and increases again.

a - Very small excitation current. The incident beam is convergent and under-focused ($\Delta h > 0$).

b - Small excitation current. The incident beam is convergent and perfectly focused. The disk of least confusion associated with the spherical aberration is observed.

c - Medium excitation current. The incident beam is divergent and over-focused ($\Delta h < 0$).

d - Medium excitation current. The incident beam is parallel.

e, f - Strong and very strong excitation currents. The incident beam is convergent and under-focused ($\Delta h > 0$).

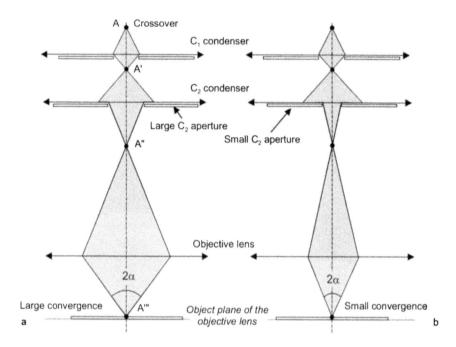

Figure VIII.9 - Effect of the size of the C_2 aperture on the convergence semi-angle α of the incident electron beam.
a - A large C_2 aperture produces a large convergence α.
b - A small C_2 aperture produces a small convergence α.

Effect of the centring of the C_2 aperture

As shown on figures VIII.10a and b, the off-axis positioning of the C_2 aperture modifies the orientation of the axis of the convergent incident beam with respect to the optic axis. This property can be used to tilt the incident beam with respect to the specimen. Nevertheless, this solution is limited to very small tilt angles since it also produces significant distortions.

VIII.2.2 - Angular scanning

Angular scanning can be used as an alternative to the LACBED technique in the "beam-rocking" technique, which will be described in the following chapter.

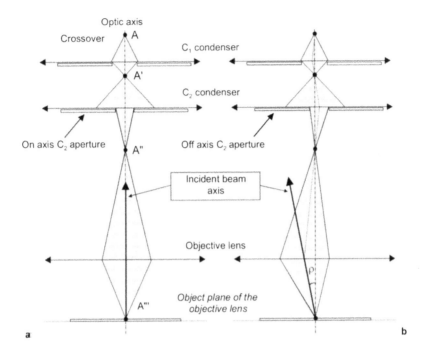

Figure VIII.10 - Effect of the centring of the C_2 aperture on the illuminating conditions.
a - The aperture is on axis. The axis of the convergent incident beam is parallel to the optic axis.
b - The aperture is off axis. The axis of the convergent incident beam is misaligned by an angle ρ with respect to the optic axis.

In this technique, a narrow parallel incident beam (obtained with the configuration described on figure VIII.7d) is rocked around a pivot point P located on the specimen (Figure VIII.11). The scanning process is carried out by means of the pre-specimen beam deflection coils. As a result, the transmitted point A''' moves, in the back focal plane of the objective lens, at the scanning rate. It gives the transmitted intensity as a function of the scanning angle. This moving point A''' can be immobilized, in the conjugate planes located in the lower parts of the microscope column, by using a second post-specimen scanning adjusted in such a way that it exactly compensates the first scanning.

This double scanning is very easy to perform with a twin lens.

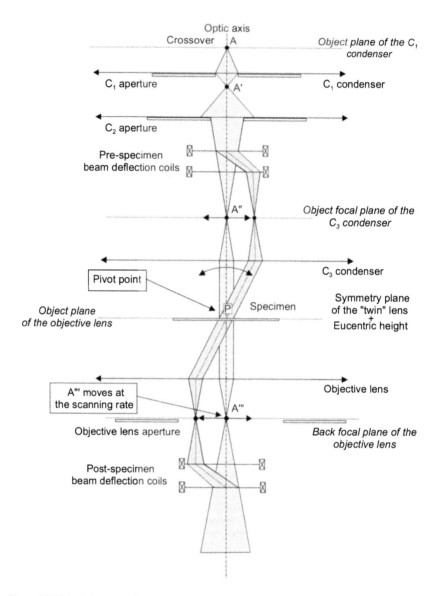

Figure VIII.11 - Schematic description of the angular scanning mode.
A narrow parallel incident beam rocks around a pivot point P located on the specimen using the pre-specimen beam deflection coils. The transmitted point A''' moves in the back focal plane of the objective lens at the scanning rate. This movement is stopped, in the conjugate planes located in the lower part of the microscope column, by the post-specimen deflection coils, which exactly compensates for the first pre-specimen scanning.

VIII.2.3 - Complete description of the formation of the image and diffraction pattern

Figures VIII.12 and VIII.13 describe the complete operation of a transmission electron microscope in the image and diffraction modes for a parallel incident beam.

In the image mode, for a bright field image, the objective aperture isolates the transmitted beam and eliminates the diffracted and scattered beams (Figure VIII.12). The microscope screen is conjugate to the object plane of the objective lens where the specimen is located.

In the diffraction mode, for selected-area diffraction, the selected-area aperture reduces the size of the diffracted area (Figure VIII.13). The screen is conjugate to the back focal plane of the objective lens where the first diffraction pattern is formed.

VIII.2.4 - Aberrations

Up to now, we have made the assumption that the magnetic lenses of the microscope are ideal and produce perfect images. In reality, magnetic lenses are far from ideal and suffer from spherical and chromatic aberrations, which affect both images and diffraction patterns. Here, we describe only the spherical aberration. The chromatic aberration will be described in chapter IX.3.

VIII.2.4.1 - Spherical aberration

In the LACBED technique, the incident beam has a very large convergence and the electrons, which depart far from the optic axis, are affected by spherical aberration. The effect of this aberration is seen on figure VIII.14, which show the formation of the image A_0' of a point object A by an ideal and a real lens.

For the ideal lens, all the rays coming from the point A focus on a single image point A'_0 (paraxial image) located in the image plane. This plane is called the **Gaussian image plane** (Figure VIII.14a).

For the real lens, the rays far from the optic axis are deviated more strongly than those that are closer to this axis. They focus on various points located between the two limiting points A'_0 (paraxial image) and A'_m (marginal image) (Figures V14b and b'). The effect is all the more important that the ray-paths are away from the optical axis. As a result, the image of a point A is no longer a point but a disk of confusion with radius r_s.

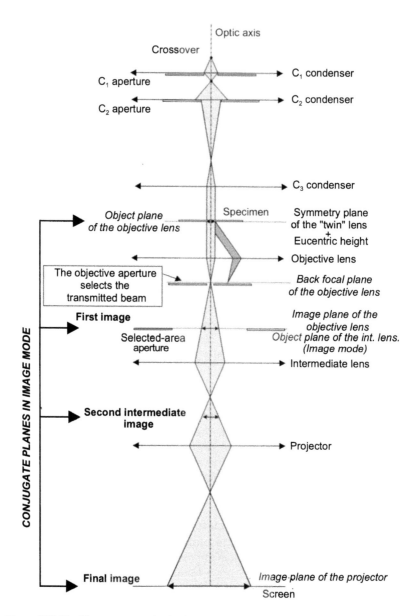

Figure VIII.12 - Electron ray-paths in the image mode for a parallel incident beam. Bright-field image under two-beam conditions. The screen is conjugate to the object and image planes of the objective lens. The objective aperture isolates the transmitted beam from the diffracted beams.

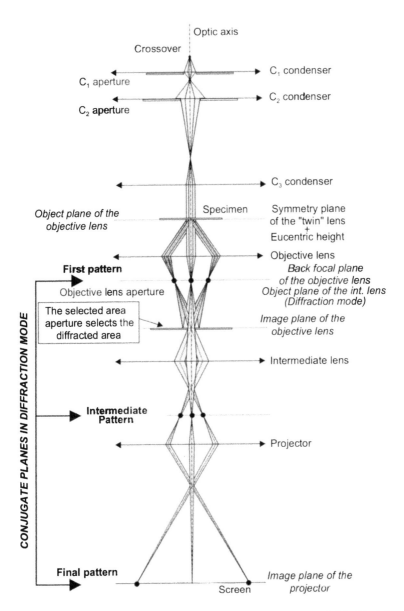

Figure VIII.13 - Electron ray-paths in the diffraction mode for a parallel incident beam. Selected-area electron diffraction. The screen is conjugate to the back focal plane of the objective lens where the first diffraction pattern is formed. The selected-area aperture defines the size of the diffracted area.

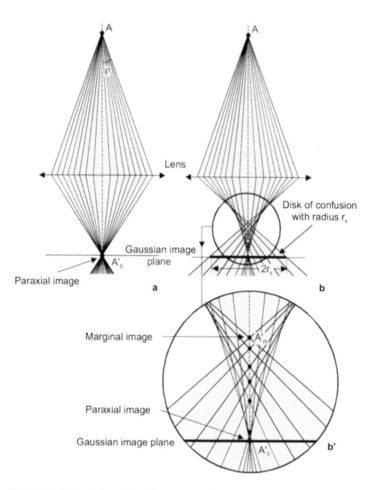

Figure VIII.14 - Spherical aberration of a magnetic lens.
a - Ideal lens. The paraxial image A'_0 of a point object A is a point located in the Gaussian image plane.
b - Real lens. The electron rays traveling far from the optic axis form images located between the paraxial point A'_0 and the marginal point A'_m. The image of a point object A becomes a disc of confusion with radius r_s in the Gaussian image plane.
b' - Enlargement of the ray-paths near the Gaussian image plane.

Figure VIII.16 gives a detailed view of the electron ray-paths as well as the appearance of the disk of confusion observed in various planes located on both sides of the Gaussian image plane (planes indicated by circled numbers).

The electron ray-paths lie on a surface of revolution consisting of an axial segment $A'_0A'_m$ and a tangential envelope. Above the Gaussian image plane, the disk of confusion consists of a bright central spot surrounded by a disk whose diameter depends on the position with respect to the Gaussian plane. For a particular value Δf, the diameter of the disk has a minimum value. It is called **disk of least confusion**.

The radius r_S of the disk of confusion depends on the beam aperture angle ρ. In the Gaussian image plane, its value is:

$$r_S = MC_s\rho^3 \tag{VIII.1}$$

C_s is the spherical aberration coefficient. Its value depends on the geometry of the magnetic lens. It is of the order of a few millimetres. M is the magnification of the lens.

The radius of the disk of least confusion is:

$$r_s = \tfrac{1}{4}MC_s\rho^3 \tag{VIII.2}$$

In these formulae, r_s is proportional to the cube of the aperture angle ρ. This characteristic is clearly seen on figures VIII.15a and b, which show two crossover images obtained with 100 and 200 µm C_2 apertures respectively. In the first case, the disk of confusion has a diameter of about 0.05 µm; in the second case, it becomes much larger and reaches 0.5 µm.

a   b

100 µm C_2 condenser   200 µm C_2 condenser

Figure VIII.15 - Effect of the size of the C_2 aperture on the crossover image observed in the Gaussian image plane.
a - 100 µm C_2 aperture. The disk of confusion has a diameter of 0.05 µm. The effect of the spherical aberration is negligible.
b - 200 µm C_2 aperture. The disk of confusion has a diameter of 0.5 µm. The effect of the spherical aberration is significant.

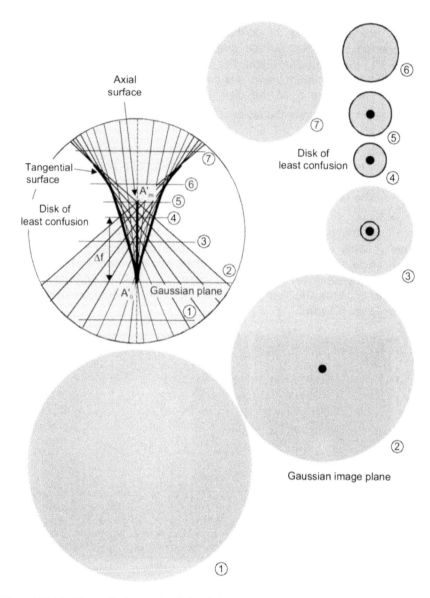

Figure VIII.16 - Theoretical aspects of the disks of confusion observed in various planes located on both sides of the Gaussian image plane. The disk of least confusion has the smallest diameter and is located at a distance Δf from the Gaussian image plane. The various planes are indicated by circled numbers.

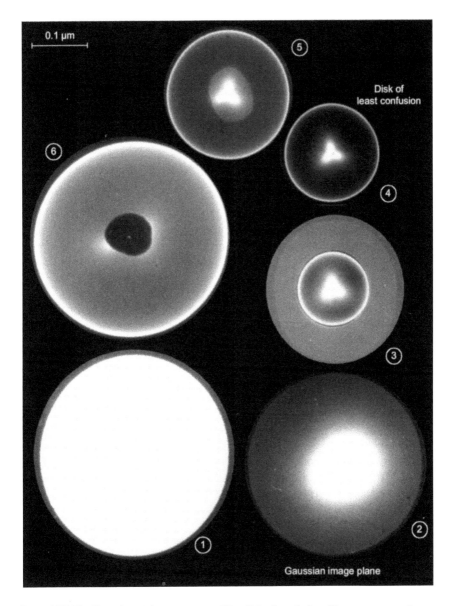

Figure VIII.17 - Experimental appearance of the disk of confusion. These crossover images were obtained by slightly defocusing the C_2 condenser on both sides of the exact focus. They are in agreement with the theoretical diagrams on figure VIII.16. The circled numbers correspond to the planes shown on figure VIII.16.

The effect of the spherical aberration is highlighted on the experimental crossover images shown on figure VIII.17. These images were obtained after slightly defocusing the C_2 condenser lens. They could also be obtained by defocusing the objective lens and hence observing object planes located on both sides of the Gaussian image plane. These images are in very good agreement with the theoretical diagrams on figure VIII.16. In particular, the disk of least confusion is well identified.

Note that the spherical aberration is mainly produced by the condenser and the objective lenses.

VIII.2.4.1.1 - Effect of the spherical aberration on Kossel and LACBED patterns

As a result of the very large convergence angles used in the Kossel and LACBED techniques, we can expect to observe strong effects connected with the spherical aberration.

- Effect on Kossel patterns

The spherical aberration has two main effects on Kossel patterns:
- it affects the superimposition of the excess and deficiency lines,
- it increases the size of the diffracted area.

- Effect of the spherical aberration on the superimposition of the excess and deficiency lines

When the diffraction lens is slightly defocused, planes located slightly above or below the back focal plane are observed and the superimposition of the excess and deficiency lines becomes imperfect. This effect is clearly visible in the experimental Kossel patterns on figure VIII.18. Moreover, according to whether the observed plane is located above or below the back focal plane, the bright excess line appears on one side or on the other side of the dark deficiency line.

This imperfect superimposition is caused by the spherical aberration and can be interpreted from figures VIII.19a and b, which show the electron ray-paths without and with spherical aberration. Without aberration, the two hkl and $\overline{hkl}$ diffraction phenomena occur at a common point E of the specimen (Figure VIII.19a) and the transmitted and diffracted ray-paths are identical. With aberration, the hkl and $\overline{hkl}$ diffraction phenomena occur at the two different points E' and E" (Figure VIII.19b) and the corresponding ray-paths are no longer common except in the back focal plane of the objective lens.

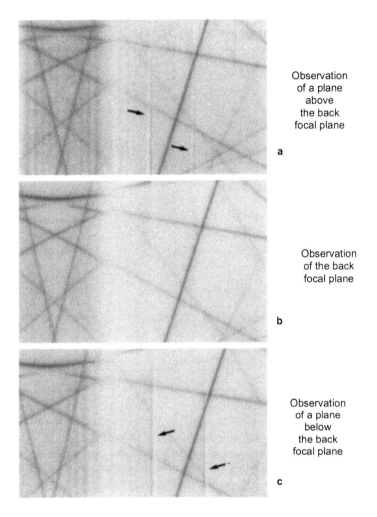

Observation
of a plane
above
the back
focal plane

a

Observation
of the back
focal plane

b

Observation
of a plane
below
the back
focal plane

c

Figure VIII.18 - Effect of the spherical aberration on the superimposition of the deficiency and excess lines present on Kossel patterns (enlargement of the central area). These Kossel patterns were obtained after slightly defocusing the diffraction lens.

a - Observation of a plane located above the back focal plane. The superimposition is imperfect. The bright excess lines, marked with an arrow, are located to the left of the corresponding dark deficiency lines.

b - Observation of the back focal plane. The superimposition is perfect in the central area of the pattern.

c - Observation of a plane located below the back focal plane. The superimposition is imperfect. The bright excess lines are now located to the right of the dark deficiency lines.

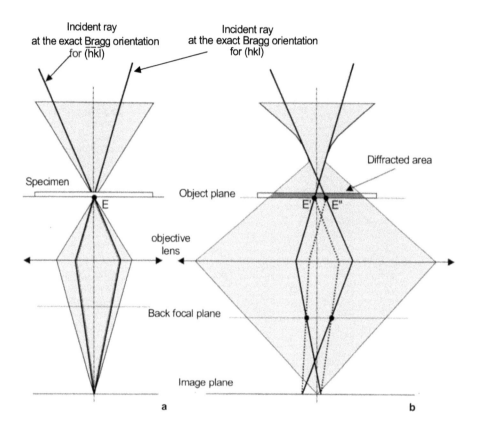

Figure VIII.19 - Effect of spherical aberration on Kossel patterns.

a - Electron ray-path without spherical aberration. The diffraction phenomena hkl and h̄k̄l̄ occur at a common point E of the specimen. The transmitted and diffracted ray-paths are identical.

b - Electron ray-paths in the presence of spherical aberration. The hkl and h̄k̄l̄ diffraction phenomena occur at two different points E' and E". The transmitted and diffracted ray-paths are common only in the back focal plane of the objective lens. The diffracted area is no longer a point.

- Effect of spherical aberration on the size of the diffracted area

The spherical aberration produces a significant widening of the diffracted area. The minimal size of this area corresponds to the disk of least confusion (Figure VIII.19b).

- Effect on LACBED patterns

A typical effect, related to the spherical aberration, is frequently observed on LACBED patterns. A dark halo, as shown on figure VIII.20, is seen.

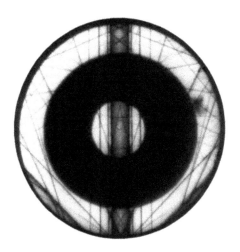

Figure VIII.20 - Effect of spherical aberration on LACBED patterns. Presence of a dark halo.

The explanation of this phenomenon was given by Vincent *et al.* [VIII.1] and can be interpreted with the help of figure VIII.21. On this figure, the C_2 condenser is set in such a way that the disk of least confusion is located roughly in the object plane of the objective lens. The first image of the disk of least confusion is formed in the image plane of the objective lens where the selected-area aperture is situated. If this aperture is slightly above the disk of least confusion, as shown on figure VIII.21, then only the rays close to the optic axis and those far from it go through the aperture. The rays located in between are eliminated and produce the dark halo observed in the back focal plane of the objective lens.

VIII.3 - Obtaining large-angle electron diffraction patterns

We first describe how to obtain Kossel patterns, before examining the practical conditions for LACBED pattern formation. The same illumination conditions are used for Kossel and LACBED patterns.

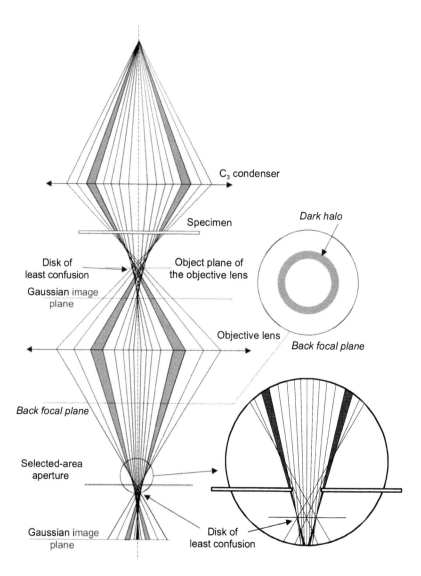

Figure VIII.21 - Effect of spherical aberration on LACBED patterns.
Presence of a dark halo. The C_2 condenser is adjusted in such a way that the image of the disk of least confusion is located slightly below the selected-area aperture. Only the rays close to or far from the optic axis go through the aperture. The rays in between are eliminated and a dark halo is formed in the back focal plane of the objective lens.

VIII.3.1 - Obtaining Kossel patterns

The description of the operating conditions is divided into two steps. The first step relates to the specimen settings and to the adjustment of the illumination conditions, which are carried out in the image mode. The second step concerns the diffraction pattern itself. It is obtained in the diffraction mode.

Figures VIII.22 and 23 give a full description of the operating conditions used to obtain a Kossel pattern both in the image and diffraction modes.

The specimen is set at the eucentric height in the microscope and is observed in the nanoprobe mode with a parallel incident beam. A magnification greater than 10000 times is advisable, since the operation of the twin lens in the nanoprobe mode can prove to be delicate at smaller magnifications.

The image of the specimen is then brought into exact focus, in order to establish the optimum operating conditions of the twin lens. These optimum conditions are shown on figures VIII.7d, VIII.8d and VIII.12.

VIII.3.1.1 - Adjustment of the illumination conditions

A probe size S is selected using the C_1 condenser. To begin with, it is preferable to use a relatively large probe size (about a few hundreds of nanometres). Subsequently, this size will be reduced in order to improve the quality of the diffraction pattern. We recall that the probe size has no effect on the diffraction pattern itself (at least with a non-distorted specimen).

The largest C_2 aperture available is selected to obtain the largest convergence angle (a diameter of 200 µm is well adapted). With some microscopes, it is also possible to work without a C_2 aperture. The polepieces of the condenser then act as a very large aperture and convergence semi-angles up to 6 or 7° can be reached.

When the C_2 condenser is used, the crossover image is formed in the object plane of the objective lens according to the configuration given in figures VIII.7b and VIII.8b. Owing to the very strong spherical aberration, a bright spot surrounded by a large disk of confusion is observed. First of all, it is sensible to observe the Gaussian image plane, which can easily be identified by means of figure VIII.17.

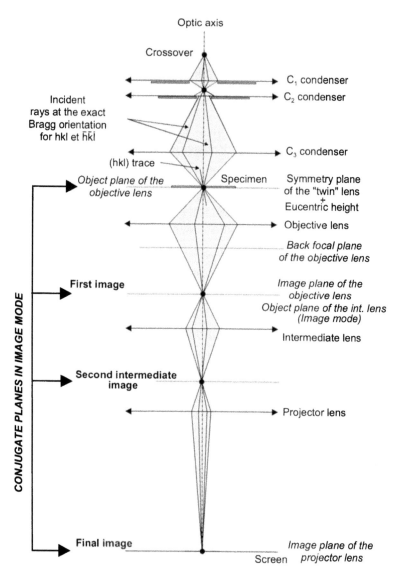

Figure VIII.22 - Kossel pattern. Complete electron ray-paths in the image mode.
The specimen is at the eucentric height and is illuminated by a convergent incident beam with convergence semi-angle α. The screen is conjugate to the object plane of the specimen.
In order to simplify this diagram, only the ray-paths at the exact Bragg orientation for the set of (hkl) lattice planes are drawn.

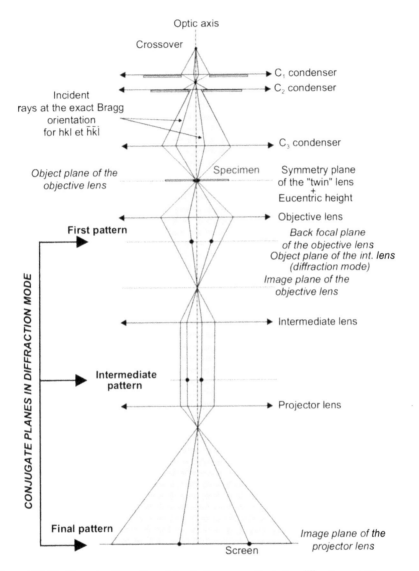

Optic axis

Crossover

→ C₁ condenser
→ C₂ condenser

Incident
rays at the exact Bragg
orientation
for hkl et h̄k̄l̄

→ C₃ condenser

Object plane of the
objective lens

Specimen

Symmetry plane
of the "twin" lens
+
Eucentric height

→ Objective lens

First pattern

Back focal plane
of the objective lens
Object plane of the int. lens
(diffraction mode)

Image plane of the
objective lens

→ Intermediate lens

Intermediate
pattern

→ Projector lens

Final pattern

Image plane of the
projector lens

Screen

CONJUGATE PLANES IN DIFFRACTION MODE

Figure VIII.23 - Kossel pattern. Complete electron ray-paths in the diffraction mode.
The specimen is located at the eucentric height and is illuminated by a convergent incident
beam with convergence semi-angle α. The screen is conjugate to the back focal plane of
the objective lens where the Kossel pattern is formed.
In order to simplify this diagram, only the ray-paths at the exact Bragg orientation for the set
of (hkl) lattice planes are drawn.

VIII.3.1.1.1 - Beam adjustments

In the Gaussian image plane, **the bright spot must be at the centre of its disk of confusion** (Figure VIII.24a). If not, (Figure VIII.24b), this means that the C_2 aperture and/or the objective lens are not aligned properly.

In order to correct this misalignment, it is necessary:
- to centre the C_2 aperture,
- to centre the objective lens.

The effects of an imperfect centring of the C_2 aperture or of the objective lens are identical on Kossel patterns. Both effects result in a misorientation of the incident beam axis and, as explained in paragraph IV.1.4, they produce a shift of the transmitted disk with respect to the centre of the screen (Figure VIII.24b'). We note that the excess and deficiency lines are not affected by this operation since they are "attached" to the specimen.

In the Gaussian image plane, **the bright spot and the disk of confusion surrounding it must have a circular shape** (Figure VIII.25a). A distorted shape as shown on figures VIII.25b and c is due to the astigmatism of the C_2 condenser and/or the objective lens. This astigmatism causes a distortion of the shape of the transmitted disk of the Kossel pattern (Figures VIII.25b' and c').

Note that excess and deficiency lines are not affected by these imperfect adjustments. This means that the Kossel patterns can be used even if the incident beam is not perfectly adjusted.

Note

Astigmatism correction is made easier if the condenser lens is slightly under-focused so that a plane located slightly above the Gaussian image plane (for example plane number 3 on figures VIII.16 and 17) is observed. Then, the centre of the disk of confusion is a bright spot surrounded by a white circle corresponding to the intersection of the plane with the tangential envelope. The use of a very small probe is also recommended so that it can be regarded as a point whose image is in agreement with the theoretical images on figure VIII.16. The patterns shown on figures VIII.24 and 25 were produced with a 2 nm probe size (size of the bright spot in the Gaussian image plane). The astigmatism of the condenser and objective lenses must be corrected independently because their effects are identical.

All these adjustments can be carried out on the specimen. However, it is advisable to carry them in the vacuum, where the image of the crossover is much clearer because it is not perturbed by inelastic scattering.

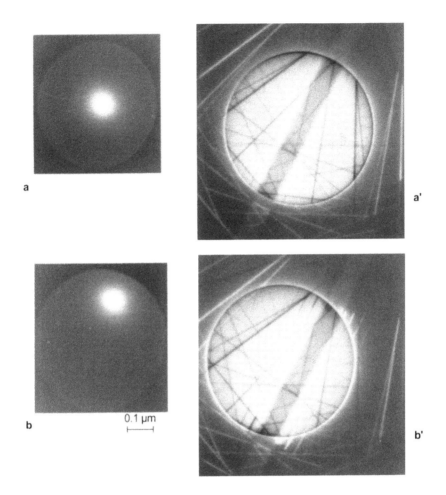

Figure VIII.24 - Kossel patterns. Effects of the centring of the C_2 aperture and/or the objective lens.

a and a' - Appearance of the Gaussian image plane and of the Kossel pattern when both the C_2 aperture and the objective lens are well centred. The bright spot is at the centre of its disk of confusion and the centre of the transmitted disk is located at the centre of the screen.

b and b' - Appearance of the Gaussian plane and of the Kossel pattern when the C_2 condenser aperture and/or the objective are imperfectly centred. The bright spot is no longer at the centre of its disk of confusion. The transmitted disk is shifted with respect to the centre of the screen. Note that the excess and deficiency lines are not disturbed by this shift.

These patterns were obtained with a 2 nm probe size (size of the bright spot in the Gaussian image plane).

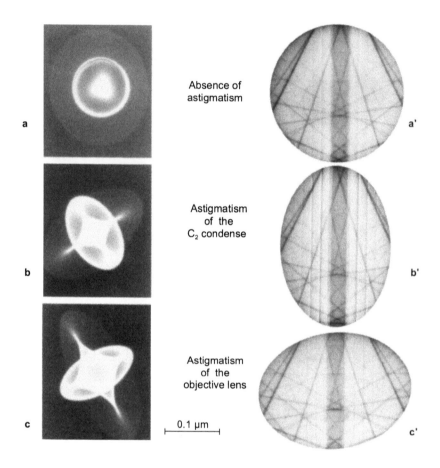

Figure VIII.25 - Appearance of the disks of confusion and of the corresponding Kossel patterns in the presence of astigmatism of the condenser and/or the objective lens. These disks are observed in a plane located slightly above the Gaussian image plane (for example, plane number 3 on figures VIII.16 and 17). The central spot of the disks of confusion is surrounded by a white circle corresponding to the intersection of the observed plane with the tangential envelope. This configuration is favourable for accurate astigmatism correction.

a and a' - Astigmatism perfectly corrected. The transmitted disk of the Kossel pattern is a circle.

b and b' - Effect of the astigmatism of the C_2 condenser lens. The transmitted disk has the shape of an ellipse.

c and c' - Effect of the astigmatism of the objective lens. It produces the same effect as that of the condenser lens. The deficiency lines are not affected by astigmatism. These patterns were obtained with a 2 nm probe size.

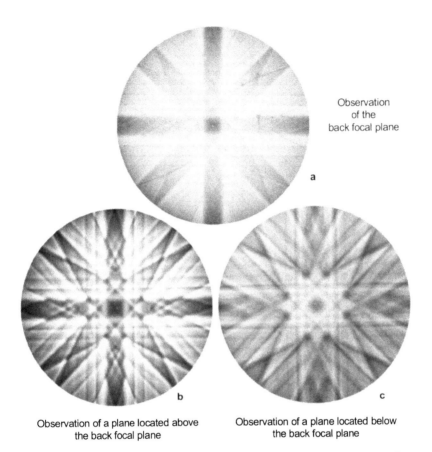

Observation
of the
back focal plane

a

b   c

Observation of a plane located above   Observation of a plane located below
the back focal plane   the back focal plane

Figure VIII.26 - Effect of the adjustment of the diffraction lens on Kossel patterns. [001] zone axis
from a silicon specimen.
a - Observation of the back focal plane. The excess and deficiency lines are superimposed in the
central area of the pattern.
b - Observation of a plane located slightly above the back focal plane. The lines are no longer
superimposed. The contrast of the pattern is improved.
c - Observation of a plane located slightly below the back focal plane. The lines are no longer
superimposed.

A very small probe size also produces more rapid contamination
of the diffracted area. This contamination can be strongly reduced by
operating at low temperatures or by flooding the specimen with electrons
for a few minutes.

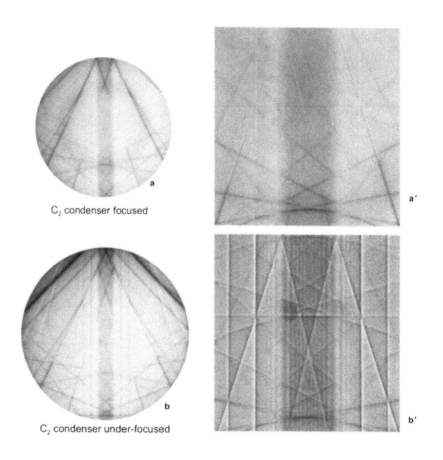

C₂ condenser focused

C₂ condenser under-focused

Figure VIII.27 - Effect of the adjustment of the C_2 condenser on Kossel patterns.
a and a' - Transmitted disk and enlargement of the central area. The condenser is focused so that the Gaussian plane is located in the object plane of the objective lens. The lines are superimposed in the back focal plane.
b and b' - Transmitted disk and enlargement of the central area of the pattern. The condenser lens is under-focused so that the disk of least confusion is located in the object plane of the objective lens. The lines are not superimposed in the back focal plane.
Note the difference in convergence between the two patterns. It results from the differences of focus of the C_2 condenser and can be interpreted using figures VIII.7a and b.

The astigmatism and the centring of both the objective and the C_2 aperture are very easily corrected with a very small probe.

VIII.3.2 - Obtaining LACBED patterns

For LACBED patterns, the illumination conditions are the same as those described for Kossel patterns.

The complete configurations in the image and diffraction modes are shown on figures VIII.28 and 29.

VIII.3.2.1 - Operating conditions in the image mode

To create a LACBED pattern, a Kossel pattern is first obtained as indicated in the previous section. Then, without changing the adjustments of the C_2 condenser and objective lenses, the specimen is raised (or lowered) in the microscope using the z adjustment of the specimen holder. This operation is carried out manually or with a motor on recent microscopes. The adjustment of the specimen height transforms the single disk of confusion into a spot pattern where each spot is a disk of confusion (see, for example, the two patterns on figures VI.6 and VI.10).

Note

For electron microscopes not equipped with a motorized holder, it is sometimes difficult to determine the upward or downward direction of the z specimen displacement. An easy way of identifying this direction consists in setting the specimen at the eucentric height, focusing its image using the objective lens and measuring the excitation current in the objective lens. The specimen height is then changed and the corresponding image is brought into focus once again. If the objective lens current decreases during this operation, this means that the specimen is above the object plane ($\Delta h > 0$). On the other hand, if it increases, the specimen is below the object plane ($\Delta h < 0$).

The smallest selected-area aperture available (2, 5 or 10 µm are good values) is inserted and centred on the transmitted beam. Since it could be difficult to centre the aperture perfectly on the transmitted beam, it is advisable to position it roughly and then to shift the incident beam with the beam shifts so that it is exactly located in the centre of the aperture. The effect of poor centring of the selected-area aperture is illustrated on figures VIII.30a and b.

VIII.3.2.2 - Operating conditions in the diffraction mode

LACBED patterns are observed by switching to the diffraction mode according to the configuration given on figure VIII.29. The screen is then conjugate to the back focal plane of the objective lens where the LACBED pattern is located.

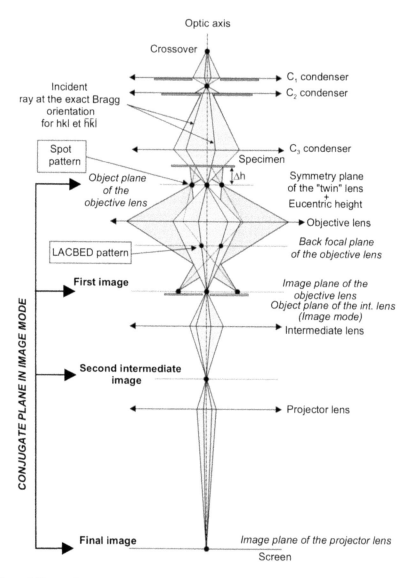

Figure VIII.28 - Bright-field LACBED pattern. Complete electron ray-paths in the image mode.
The condenser is focused on the object plane of the objective lens. The specimen is raised (or lowered) by the distance Δh from this plane. The screen is conjugate to the object plane where the spot pattern is located. The selected-area aperture selects the transmitted beam from the diffracted beams.

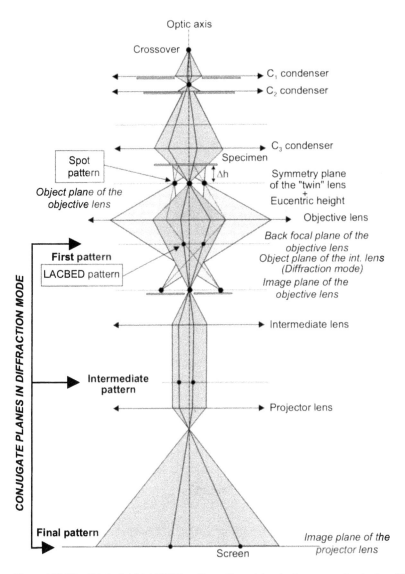

Figure VIII.29 - Bright-field LACBED pattern. Complete electron ray-paths in the diffraction mode.
The condenser lens is focused in the object plane of the objective lens. The specimen is raised (or lowered) by the distance Δh from this plane. The screen is conjugate to the back focal plane of the objective lens where the LACBED pattern is formed. The selected-area aperture selects the transmitted beam from the diffracted beams.

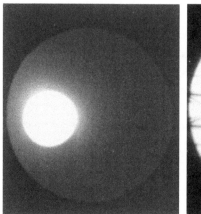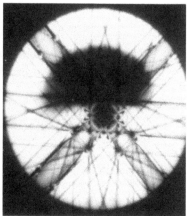

a   b

Figure VIII.30 - Effect of poor centring of the selected-area aperture on LACBED patterns.
a - In the image mode. The crossover image is not centred with respect to the selected-area aperture.
b - In the diffraction mode. A dark halo appears on the LACBED pattern.

VIII.3.2.2.1 - Dark-field LACBED pattern

A diffracted beam must be isolated using the selected-area aperture to obtain a dark-field LACBED pattern. Two solutions can be used.

- The first solution consists in positioning the selected-area aperture around the corresponding diffracted spot of the spot pattern (Figure VIII.31a).

- The second solution consists in shifting the spot pattern, using the beam shifts to place the chosen spot in the centre of the selected-area aperture (Figure VIII.31b).

These two solutions give similar results.

VIII.3.2.3 - Quality improvements of LACBED patterns

The main adjustments required to improve the quality of LACBED patterns are:

- the perfect centring of the C_2 aperture and of the objective lens. The effects of poor centring are shown on figure VIII.32a. They do not affect the deficiency lines.

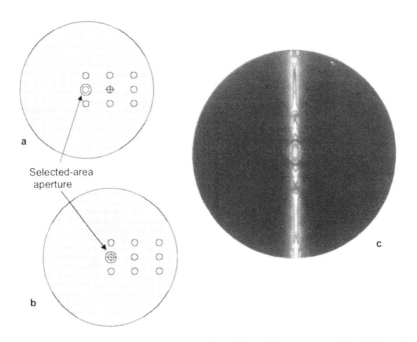

Figure VIII.31 - Dark-field LACBED pattern. Description of the two possible adjustments of the selected-area aperture.
a - The selected-area aperture is centred on the chosen diffracted spot of the spot pattern.
b - The spot pattern is shifted using the beam shifts in order to place the selected diffracted spot on the optic axis of the microscope. The selected-area aperture is then centred on this optic axis.
c - Example of dark-field LACBED pattern.

- the perfect correction of the astigmatism of the C_2 condenser and/or of the objective lens. The effects of poor correction are seen on figure VIII.32b. As before, they do not affect the individual features of the pattern.

- the optimum adjustment of the C_2 condenser. If this lens is under-focused, the dark halo described in paragraph VIII.2.4.1.1 is observed. The optimum adjustment of the C_2 condenser is obtained when **the disk of confusion fills the selected-area aperture.** It is under these conditions that the largest convergence is obtained, as shown on figure VIII.33b. The C_2 condenser is thus slightly under-focused (Figure VIII.33c). If the C_2 condenser is exactly focused, the Gaussian image plane is conjugate to the selected-area aperture and the convergence decreases (Figure VIII.33a).

The beam convergence can be increased with a large selected-area aperture. However, the pattern quality becomes poor since the inelastic filtering is less effective (see paragraph VI.6.8). The effect of 10 and 50 μm selected-area apertures on the convergence is shown on figure VIII.34. The convergence semi-angle increases from 2.5° to 3.3°.

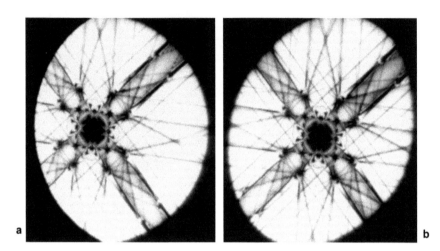

Figure VIII.32 - LACBED pattern.
a - Effect of poor centring of the C_2 aperture and/or the objective lens.
b - Effect of poor correction of the astigmatism of the C_2 and/or the objective lens.
Note that the individual features in these two patterns are not affected by these poor alignments.

- The selection of the probe size S. We indicated, in chapters VI.5 and VI.6.6, that the probe size has no effect on the line pattern but has a very strong effect on the resolution of the superimposed shadow image. This property should be taken into account whenever the superimposed image is to be observed. It is especially the case for the analysis of crystal defects.

- the perfect adjustment of the diffraction lens. The effect of a poor adjustment of the diffraction lens is illustrated on figure VIII.35. It results in a change of the magnification of the diffraction pattern. The sharpest Bragg lines are observed only in the back focal plane (Figure VIII.35b'). The pattern on figure VIII.35b, obtained with no selected-area aperture, can be used to locate the back focal plane accurately. It corresponds to the best superimposition of the excess and deficiency lines.

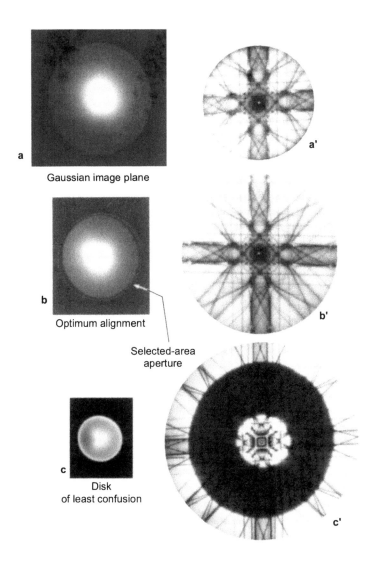

a
Gaussian image plane

a'

b
Optimum alignment

b'

Selected-area
aperture

c
Disk
of least confusion

c'

Figure VIII.33 - Optimum adjustment of the C_2 condenser. Disk of confusion and LACBED patterns.
a and a' - The C_2 condenser is focused on the Gaussian plane.
b and b' - The C_2 condenser is under-focused so that the disk of confusion fills the selected-area aperture. The convergence is optimum.
c and c' - The C_2 condenser is under-focused on the disk of least confusion. The convergence is larger but a dark halo disturbs the pattern.

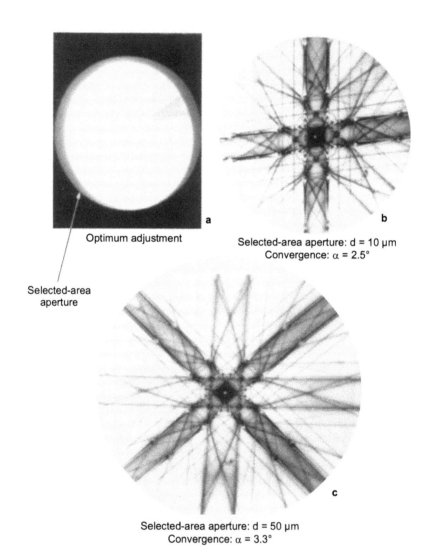

a
Optimum adjustment

Selected-area
aperture

b
Selected-area aperture: d = 10 µm
Convergence: α = 2.5°

c
Selected-area aperture: d = 50 µm
Convergence: α = 3.3°

Figure VIII.34 - Effect of the size of the selected-area aperture on the convergence of LACBED patterns.
a - The largest convergence is obtained when the disk of confusion entirely fills the selected-area aperture.
b -10 µm selected area aperture. The convergence semi-angle is 2.5°.
c - 50 µm selected area aperture. The convergence semi-angle is 3.3°. It is larger but the filtering of the inelastic electrons is less effective.

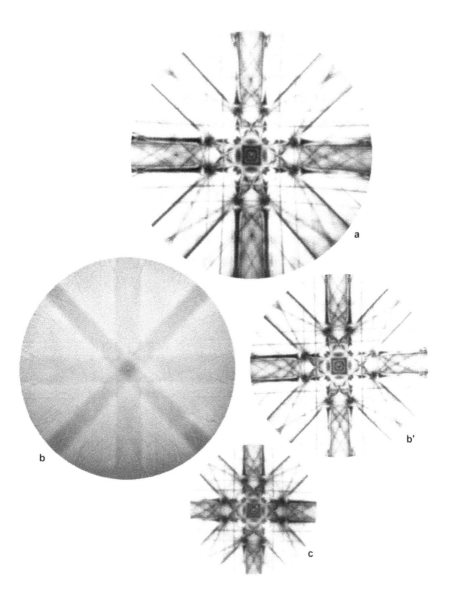

Figure VIII.35 - Effect of a misadjustment of the diffraction lens on LACBED patterns.
a - Observation of a plane located below the back focal plane.
b and b' - Observation of the back focal plane without and with a selected-area aperture. The superimposition of the lines is perfect in the central area of the pattern b.
c - Observation of a plane located above the focal plane. The magnification of the LACBED pattern is modified.

- the choice of the specimen height Δh. In chapter VI.2, we indicated that the height of the specimen has no effect on the line pattern. This is only true if the specimen is not distorted. In the case of a strongly distorted specimen, the height Δh must be reduced as far as possible in order to decrease the size of the illuminated area and, consequently, its deformations. However, the specimen height must be sufficiently large for the transmitted and diffracted beams to be selected by the selected-area aperture. Of course, a very small selected-area aperture produces a favourable effect. In chapter XI, which is concerned with applications, we shall see that the correct setting of the specimen height is particularly important for the analysis of crystal defects.

Chapter IX
LACBED variants

Several variants of the LACBED technique have been proposed. Some of these are designed to avoid the disadvantages and limitations of the LACBED technique, others are intended to improve specific properties. These are essentially:

- the defocus CBED technique,
- the eucentric LACBED technique,
- the CBIM (Convergent-Beam Imaging) technique,
- the beam-rocking technique, also called SACP (Selected-Area Channelling Pattern),
- the specimen-rocking technique,
- the bright + dark field LACBED technique,
- the montage of CBED patterns.

These variants can be sorted into several categories.

- Parallel or serial techniques. In the first case, the whole diffraction pattern is recorded simultaneously with a stationary incident beam in conventional transmission electron microscopy. In the second case, the pattern is recorded sequentially with an incident beam that scans the specimen in an angular manner. The scanning mode of the transmission electron microscope is an example of this. Thanks to the scanning process, the diffraction pattern is directly digitised. The beam-rocking technique, the specimen-rocking technique and the montage of CBED patterns belong to this category.
- Defocused techniques. In this case, the incident beam is focused above or below the specimen. We recall that information both on the direct and reciprocal spaces is available simultaneously with such a defocused incident beam. This important property is mainly used for the analysis of crystal defects. The defocus CBED, the eucentric LACBED and the CBIM techniques belong to this category.

- Techniques that favour the image to the detriment of the diffraction pattern. This is the case with the CBIM technique.

- Techniques that maintain the specimen at the eucentric height. Setting the specimen at the eucentric height can prove to be a major advantage when the specimen is tilted. For example, when a dislocation is tilted from one orientation to another, its contrast may disappear. If the specimen is not at the eucentric height, the dislocation can be lost. This problem can be solved by using the eucentric LACBED, the CBIM or the defocus CBED techniques.

In this chapter, the LACBED variants will be classified into parallel and serial techniques.

IX.1 - Parallel techniques

IX.1.1 - The eucentric LACBED technique

In the usual LACBED technique, the specimen is raised ($\Delta h > 0$) (Figure IX.1a) or lowered ($\Delta h < 0$) (Figure IX.1b) with respect to the object plane of the objective lens where the convergent incident beam is focused. Thus, the specimen is not at the eucentric height since this height usually corresponds to the object plane of the objective lens.

A second solution consists in positioning the specimen at the eucentric height and focusing the beam below (Figure IX.1c) or above it (Figure IX.1d). We have indicated, in the previous chapter, how to perform this operation using the C_2 condenser (see figures VIII.7 and 8).

With respect to the specimen, these two experimental situations are identical. Nevertheless, they are different with respect to the microscope for the following reasons.

With the conventional LACBED technique (Figures VIII.28, 29 and IX.1), the spot pattern (real or virtual according to the sign of Δh) is located both in the object plane and in its conjugate image plane. The corresponding LACBED pattern is located in the back focal plane of the objective lens. In the image mode, the image plane is conjugate to the screen (Figure VIII.28) where a magnified spot pattern is observed. In the diffraction mode, the back focal plane of the objective lens is conjugate to the screen, where a magnified LACBED pattern is seen (Figure VIII.29).

With the eucentric LACBED technique, the C_2 condenser is set so that the crossover image is formed below (or above) the specimen. As a result, the magnified image of the spot pattern is not located in the image plane but below it (Figure IX.2a) and the transmitted and diffracted beams can no longer be separated with the selected-area aperture located in this image plane.

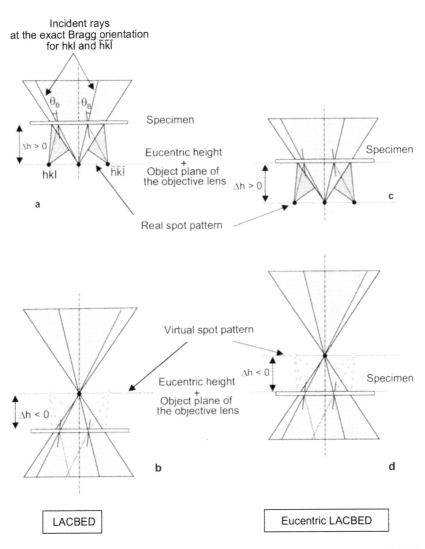

Figure IX.1 - Illumination conditions used in the conventional and eucentric LACBED techniques.

a, b - Conventional LACBED technique. The specimen is raised ($\Delta h > 0$) or lowered ($\Delta h < 0$) with respect to the eucentric height. The spot pattern (real or virtual according the sign of Δh) is located at the eucentric height.

c, d - Eucentric LACBED technique. The specimen is located at the eucentric height and the spot pattern is located below or above this eucentric height.

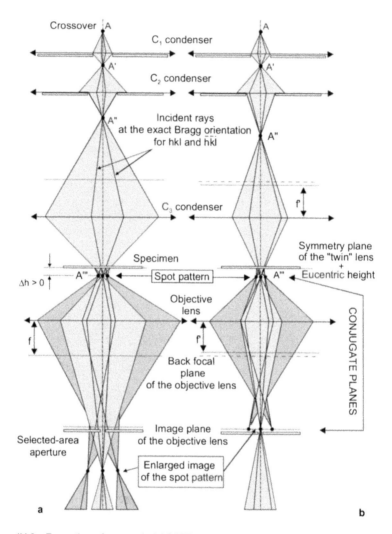

Figure IX.2 - Formation of eucentric LACBED patterns. The C_2 condenser is under-focused (or over-focused) in order to produce a crossover image A''' located below (or above) the eucentric height.

a - Normal adjustment of the objective lens. The magnified image of the spot pattern is formed below the selected-area aperture. The transmitted and diffracted beams cannot be separated.

b - The objective lens is adjusted so that the magnified image of the spot pattern is formed in the plane of the selected-area aperture. The transmitted beams can be isolated and separated from the diffracted ones. Since this modification produces a slight change of the C_3 focal length, an adjustment of the C_2 condenser is also required.

To separate the transmitted beam from the diffracted beams with the selected-area aperture, the objective lens must be readjusted (Figure IX.2b). This modification of the objective lens alters its focal length. This means that the diffraction lens must also be slightly readjusted, in the diffraction mode, so that the back focal plane of the objective lens and the screen again become conjugate.

IX.1.1.1 - Obtaining an eucentric LACBED pattern

The experimental technique shown on figure IX.2 consists in defocusing the C_2 condenser followed by a further defocusing of the objective lens. In practice, this is rather difficult with a twin lens because a change of the objective lens also involves a change of the C_3 condenser. A much simpler solution consists in defocusing the objective lens before defocusing the C_2 condenser. We describe this solution.

IX.1.1.1.1 - In the image mode

The technique involves the following steps.

- The specimen is set at the eucentric height and is observed with a parallel incident beam.
- The image is brought into focus (Figure IX.3a), which means that the object plane is located at the eucentric height.
- The objective lens is slightly defocused in order to raise or to lower its object plane with respect to the eucentric height. The image of the specimen goes out of focus (Figure IX.3b).
- The incident beam is focused in this new object plane using the C_2 condenser. A spot pattern is then observed (Figure.IX.3c).
- The selected-area aperture is inserted to select the transmitted beam or a diffracted beam.

IX.1.1.1.2 - In the diffraction mode

In order to observe the eucentric LACBED pattern, the microscope is switched to the diffraction mode (Figure IX.3d). A slight readjustment of the diffraction lens might be necessary.

In order to change the Δh value, the objective lens is slightly defocused. This operation has also an effect on the convergence semi-angle.

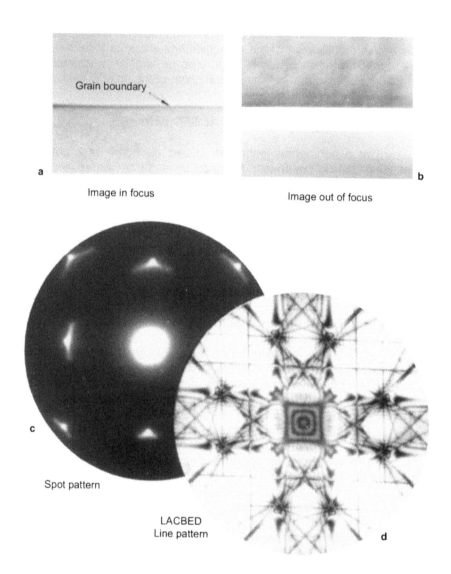

Image in focus

Image out of focus

Spot pattern

LACBED
Line pattern

Figure IX.3 - Formation of eucentric LACBED patterns.
a - The specimen, located at the eucentric height, is observed with a parallel incident beam. The objective lens is adjusted in order to bring the image into focus.
b - The objective lens is readjusted. The image of the specimen becomes out of focus.
c - The C_2 condenser is focused. The crossover image becomes a spot pattern.
d - The microscope is switched to the diffraction mode and the selected-area aperture is inserted. The eucentric LACBED pattern is observed in the back focal plane.

IX.1.1.2. - Improvement of the quality of eucentric LACBED patterns

The experimental parameters used to improve the quality of the eucentric LACBED patterns are the same as those reported for the LACBED patterns. They are described in paragraph VIII.3.2.3.

In conclusion, the main advantage of the eucentric LACBED technique concerns the specimen, which remains located at the eucentric height. The disadvantage of this technique is that the microscope is not operating with its optimal performance. Despite this problem, the eucentric LACBED patterns display an excellent quality provided the value of Δh is not too large. The example given on figure IX.3d illustrates this property.

IX.1.2 - The CBIM technique

CBIM means Convergent Beam Imaging. For this technique, proposed by Humphreys *et. al.* [IX.1, 2], the illumination conditions of the specimen are the same as those described for the "eucentric" LACBED technique.
- The specimen is located at the eucentric height (Figure IX.4).
- The incident beam is focused below or above the specimen (the C_2 condenser is over-focused or under-focused).
- The objective lens is normally adjusted so that its object plane coincides with the eucentric height. Unlike LACBED patterns, where the back focal plane of the objective lens is conjugate to the screen, the image plane of the objective lens is conjugate to the screen for CBIM patterns (Figure IX.4). Thus, all observations are made in the image mode.
In this image plane, the image of the illuminated area is in focus (since the object and the image planes of the objective lens are conjugate), but the line pattern is not (it is only in focus in the back focal of the objective lens). Nevertheless, according to figure IX.5b, if a very small probe size S is used, the broadening of the deficiency and excess lines remains moderate and the pattern displays a good enough quality.
Like LACBED patterns, the CBIM patterns provide information on both the direct space (the image of the illuminated area) and the reciprocal space (the line pattern). For LACBED patterns, the line pattern is in focus and the image is out of focus. The opposite situation occurs for CBIM patterns where the image is in focus and the diffraction pattern is not.

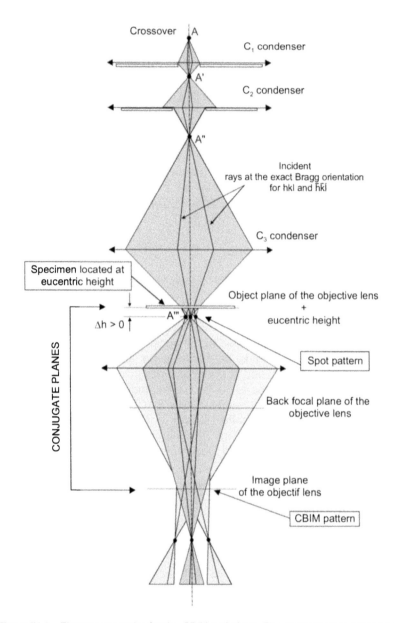

Figure IX.4 - Electron ray-paths for the CBIM technique. The convergent incident beam is under-focused (or over-focused) in such a way that the crossover image A'" is located below (or above) the eucentric height. The CBIM pattern is observed in the image plane of the objective lens, which is conjugate to the object plane.

IX.1.2.1 - Comparison of the LACBED and CBIM techniques

The LACBED and CBIM techniques are compared on Figure IX.5, where the diffraction phenomena generated by a single set of (hkl) lattice planes are described [IX.3].

For the LACBED technique (Figure IX.5a), the hkl deficiency line and the $\overline{hkl}$ excess line as well as the $\overline{hkl}$ deficiency line and the hkl excess lines are superimposed in the back focal plane of the objective lens. They can be separated by the selected-area aperture to form either a bright- or a dark-field pattern.

For the CBIM technique (Figure IX.5b), the same superimposition occurs in the back focal plane. In the image plane, however the hkl excess and deficiency lines on one hand and the $\overline{hkl}$ deficiency and excess lines on the other hand are superimposed. The experimental CBIM pattern on figure IX.7b shows that this superimposition is imperfect. The strong spherical aberration is the cause of this and the effect may be interpreted by using figure IX.6. The diffracted and transmitted beams coming from a point A of the specimen have different ray-paths and thus suffer different spherical aberration.

The superimposition of the excess and deficiency lines can be removed by inserting the objective aperture. Figure IX.5b shows that this operation has the great disadvantage of strongly reducing the convergence of the CBIM patterns.

IX.1.2.2 - Obtaining CBIM patterns

All operations are done in the image mode according to the following steps.
- The specimen is located at the eucentric height. Its image, observed under the usual conditions, is brought into focus (Figure IX.7a).

- The C_2 condenser is adjusted so that the convergent incident beam is focused above or below the specimen. More or less superimposed Bragg contours appear (Figure IX.7b).

- The objective aperture is inserted in order to remove the superimposition of the Bragg contours (Figure IX.7c).

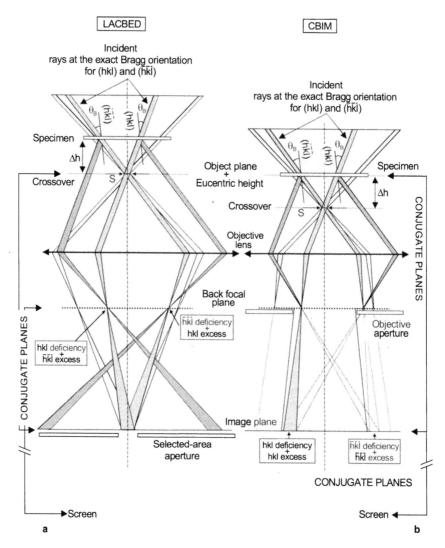

Figure IX.5 - Comparison of the LACBED and CBIM techniques.
a - LACBED technique. The back focal plane of the objective lens is conjugate to the screen where the hkl deficiency and the h̄k̄l̄ excess lines are superimposed. The selected-area aperture removes the superimposition by eliminating the diffracted beams.
b - CBIM technique. The image plane of the objective lens is conjugate to the screen. In this screen, the hkl deficiency and excess lines as well as the h̄k̄l̄ deficiency and excess lines are superimposed. The objective aperture removes this superimposition by eliminating the diffracted beams.

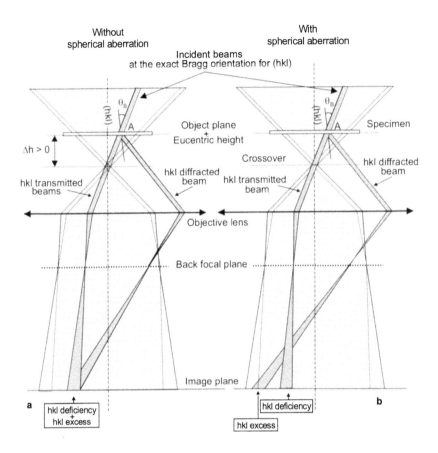

Figure IX.6 - Effect of the spherical aberration on the superimposition of the excess and deficiency lines present on CBIM patterns.
a - Electron ray-path for a perfect objective lens. The hkl excess and deficiency lines are perfectly superimposed in the image plane.
b - Electron ray-path for an objective lens with spherical aberration. The hkl diffracted beam far from the optic axis is more deviated than the hkl transmitted beam closer to this axis. As a result, the hkl excess and deficiency lines are not exactly superimposed in the image plane.

Note
CBIM patterns are observed in the image mode. The deficiency and excess lines are the images of the specimen loci at the exact Bragg orientation. For this reason, these lines are called **Bragg contours.** They are formed in the same way as the bend contours obtained with a parallel incident beam illuminating a bent specimen (see paragraph VI.14). The term **Bragg lines** will be reserved for the excess and deficiency lines observed in the diffraction mode in the back focal plane of the objective lens. This is the case for the lines present on CBED, Kossel and LACBED patterns.

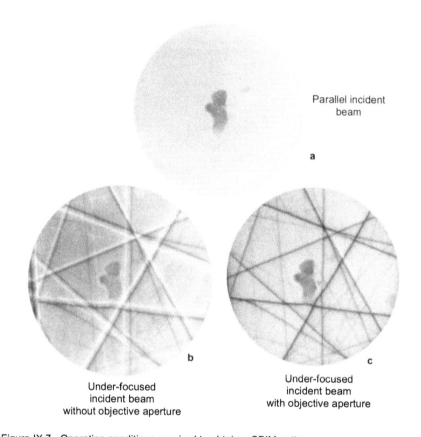

Parallel incident
beam

a

Under-focused
incident beam
without objective aperture

b

Under-focused
incident beam
with objective aperture

c

Figure IX.7 - Operating conditions required to obtain a CBIM pattern.
a - The specimen, located at the eucentric height, is illuminated with a parallel incident beam. Its image is brought into focus.
b - The incident beam is under-focused (or over-focused) with the C_2 condenser. Excess and deficiency Bragg contours appear. As a result of spherical aberration, they are not perfectly superimposed.
c - The objective aperture is inserted in order to remove the superimposition of the excess and deficiency Bragg contours.

IX.1.2.3 - Effects of the experimental parameters on CBIM patterns

IX.1.2.3.1 - Effect of the focus of the C_2 condenser

Figure IX.8 shows the appearance of the CBIM patterns observed with various excitations of the C_2 condenser (these experimental conditions are identical with those shown on figure VIII.8).

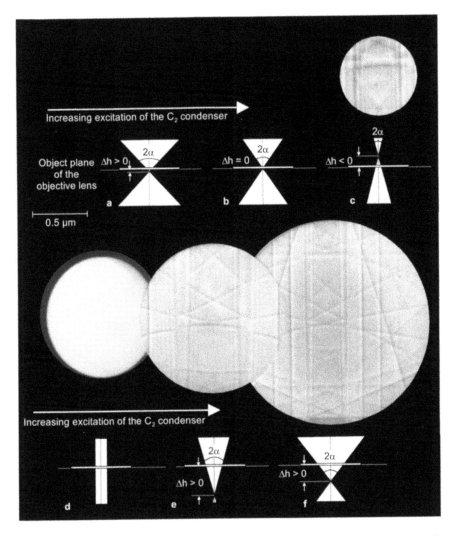

Figure IX.8 - Appearance of CBIM patterns as a function of the excitation of the C_2 condenser.

a, b - Very small excitation currents. The convergent incident beam is slightly under-focused in a and focused in b. The CBIM patterns are too small to be used.

c - Small excitation current. The incident beam is divergent. The size of the pattern and the convergence semi-angle ($\alpha = 0,4°$) are small.

d - Medium excitation current. The incident beam is parallel. Bragg contours are absent (if the specimen is perfectly flat).

e, f - High and very high excitation currents. The incident beam is convergent and strongly over-focused. The size of the pattern and the convergence semi-angle ($\alpha = 0.6°$ at e and $1.7°$ at f) are very large.

When the excitation current of the C_2 condenser increases, the size of the illuminated area decreases until a minimum is obtained for the perfect focus (Figure IX.8b), after which it increases again. Only the high operating currents, located on both sides of the parallel beam illumination, are used (c to f on figure IX.8).

The low excitation currents (a and b on figure IX.8) produce CBIM patterns that are too small. The convergence semi-angle becomes very significant for high operating currents. The maximum value obtained with the twin lens of a Philips CM30 microscope is about 6°, which is considerable (Figure IX.9). In this case, the CBIM method might well be named LACBIM (Large-Angle Convergent-Beam Imaging).

Figure IX.9 - CBIM pattern obtained with a very high excitation current of the C_2 condenser on a Philips CM30 microscope. The convergence semi-angle α reaches 6°. Silicon specimen with orientation close to the [001] zone axis.

The magnification of the Bragg contours on a CBIM pattern depends on the defocus value Δh. On the other hand, the magnification of the image of the illuminated area is not affected by Δh. This property is the opposite of that observed for LACBED patterns where a variation of Δh has an influence on the image but not on the diffraction pattern.

Note that under or over-focusing the C_2 condenser (positive or negative values for Δh), results in a 180° rotation of the diffraction pattern with respect to the image (Figure IX.10). Let us recall that for LACBED patterns, the image undergoes such a rotation but the diffraction pattern does not.

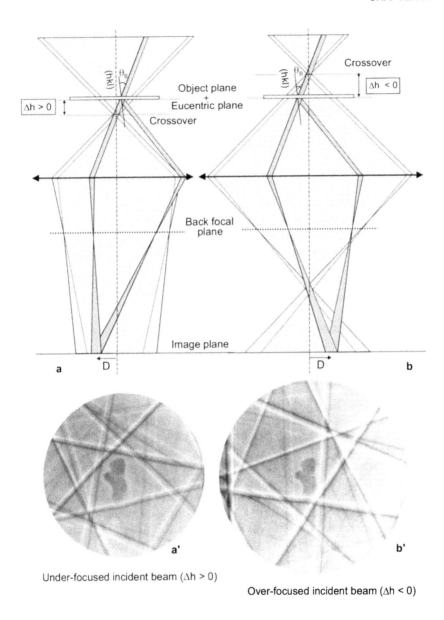

Under-focused incident beam (Δh > 0)

Over-focused incident beam (Δh < 0)

Figure IX.10 - Effect of over- and under-focusing of the incident beam on CBIM patterns.
a, a' - Under-focused incident beam (Δh > 0).
b - Over-focused incident beam (Δh < 0). The CBIM pattern undergoes a 180° rotation with respect to the previous case. Note that the image remains unchanged.

IX.1.2.3.2 - Effect of the probe size S

As shown on figure IX.11, the probe size S has a direct effect on the width L of the Bragg contours. The smaller the spot size, the sharper the Bragg contours. For this reason, the nanoprobe mode, which is the only operating mode of the microscope able to produce very small probe sizes (from 1 to 100 nm) must be used. Obviously, the probe size has no effect on the image of the illuminated area.

IX.1.2.3.3 - Effect of the objective aperture

The objective aperture has three effects.
- It removes the superimposition of the excess and deficiency lines (figures IX.7b and IX.12b).
- It filters the inelastic electrons and improves the quality of CBIM patterns. This property will be developed in paragraph IX.3.
- It reduces the convergence semi-angle. This unfavourable effect is illustrated on figures IX.12a and b.

IX.1.2.4 - Advantages and disadvantages of the CBIM technique

The CBIM technique is very simple to perform. Starting from a normal observation with a parallel incident beam, it is only necessary to increase or to decrease the excitation of the C_2 condenser.

Since the CBIM technique is a defocused technique, it is very well adapted to the study of crystal defects. Moreover, the position of the specimen at the eucentric height is very useful for the analysis of poorly contrasted defects.

Nevertheless, the quality of the CBIM patterns is definitely worse than that of the LACBED patterns, which is excellent. To a large extent, this poor quality is due to inefficient filtering of the inelastic electrons by the objective aperture. In order to remove more inelastic electrons, this aperture would have to be very small, but then the convergence of the CBIM pattern becomes too small and the pattern is no longer useful.

In paragraph IX.3.1.2.3, we show how the quality of CBIM patterns can be improved by using an energy filter.

We recall that, for LACBED patterns, the selected-area aperture removes most of the inelastic electrons.

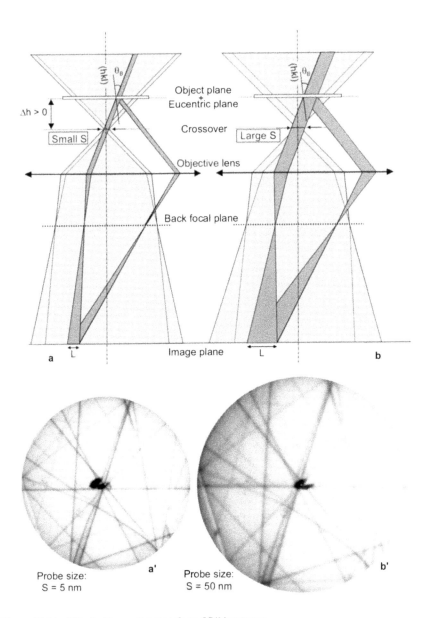

Figure IX.11 - Effect of the probe size S on CBIM patterns.
a, a' - Small probe size (5 nm for the experimental pattern). The deficiency lines are sharp.
b, b' - Large probe size (80 nm for the experimental pattern). A significant broadening of the excess and deficiency lines is observed.
Note that the probe size has no influence on the image.

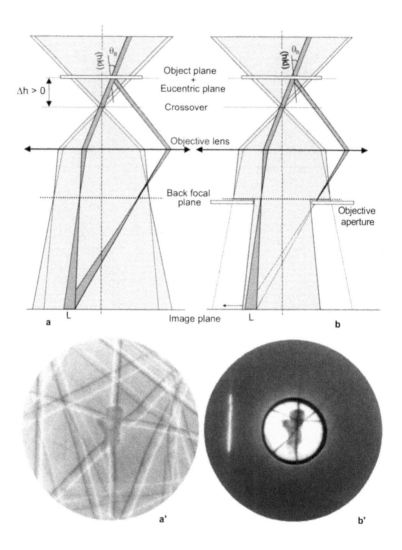

Figure IX.12 - Effect of the objective aperture on CBIM patterns.
a, a' - Electron ray-paths and CBIM pattern in the absence of objective aperture. The excess and deficiency lines are imperfectly superimposed.
b, b' - Electron ray-paths and CBIM pattern in the presence of an objective aperture. The aperture prevents the superimposition of the excess and deficiency lines inside the transmitted disk. It also produces a strong reduction of the convergence semi-angle and an improvement of the quality of the pattern related to the angular filtering of the inelastic electrons.

IX.1.3 - The defocus CBED technique

This technique, proposed by Tanaka *et al.* [IX.4], favours the "defocus" aspect in order to obtain information on both the direct and reciprocal spaces.

A conventional CBED pattern is first obtained with a convergence selected in such a way that the transmitted and diffracted disks do not overlap (Figure IX.13a).

The excitation of the C_2 condenser is then slightly modified in order to focus the incident beam above or below the specimen (Figures IX.13b and c). The image a'b' of the illuminated area ab of the specimen appears inside the transmitted disk and inside each diffracted disk (Figures IX.14 and 15).

IX.1.3.1 - Obtaining a defocus CBED pattern.

The specimen is located at the eucentric height and its image is brought into focus. The incident beam is focused on the specimen (the image of the crossover is formed on the specimen). In the diffraction mode, a conventional CBED pattern is observed in the back focal plane of the objective lens, (Figure IX.13a). The C_2 condenser is then slightly modified in order to form the crossover image above ($\Delta h < 0$) or below ($\Delta h > 0$) the specimen. During this operation, the diffraction pattern is not modified but the shadow image of the illuminated specimen area appears in each disk. Depending on whether the incident beam is under or over-focused, the shadow image is or not rotated by 180° (Figures IX.13 a, c).

Note
For CBIM patterns, the illumination conditions are located on both sides of the parallel beam condition. For the defocus CBED technique, they are located on both sides of the perfect focus condition. In view of the small convergence angles involved, the microprobe mode is well adapted to this type of patterns.

IX.1.3.2 - Main experimental parameters

The two main experimental parameters involved in this technique are the defocus of the C_2 condenser and the probe size. The defocus of the C_2 condenser changes the Δh value and therefore the magnification of the shadow image. The probe size has a direct effect on the resolution of the shadow image.

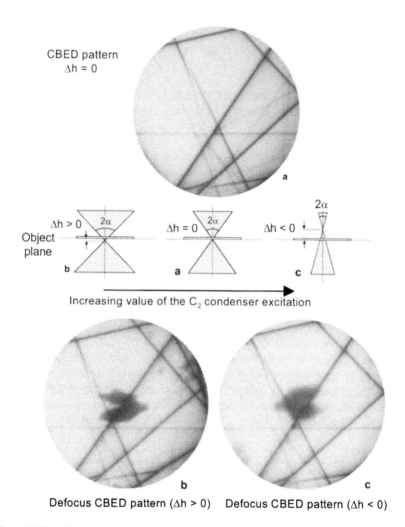

CBED pattern
Δh = 0

Object plane

Increasing value of the C_2 condenser excitation

Defocus CBED pattern (Δh > 0)   Defocus CBED pattern (Δh < 0)

Figure IX.13 - Illumination conditions for the defocus CBED technique. Observation of the transmitted disk.

a - The incident beam is focused on the specimen (Δh = 0) in order to obtain a conventional CBED pattern.

b - The incident beam is under-focused (Δh > 0). The image of the illuminated area is superimposed on the pattern.

c - The incident beam is over-focused (Δh < 0). The image of the illuminated area of the specimen is rotated by 180° with respect to the pattern b. The image of the illuminated area remains unchanged.

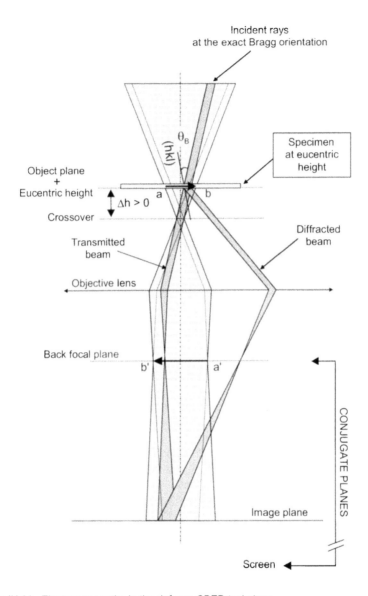

Figure IX.14 - Electron ray-paths in the defocus CBED technique.
The specimen is located at the eucentric height. The crossover image is located below (or above) the specimen with the help of the C_2 condenser. The pattern is observed in the back focal plane. The shadow image a'b' of the illuminated area ab of the specimen is superimposed onto the line pattern.

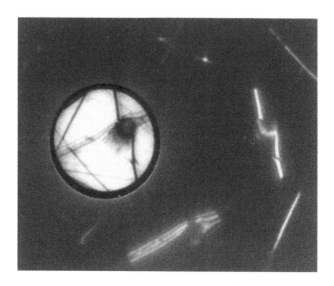

Figure IX.15 - Defocus CBED pattern. It is composed of a transmitted disk and several diffracted disks. The shadow image of the illuminated area of the specimen is observed in each of these disks.

IX.1.4 - The bright + dark field LACBED method

This technique, proposed by Terauchi and Tanaka [IX.5], gives LACBED patterns containing simultaneously the transmitted and the diffracted disks without any overlap.

Consider figure IX.16. At the level of the plane L, the transmitted and diffracted beams are separated and do not overlap. In order to obtain a bright + dark field LACBED pattern, the intermediate lens should be adjusted so that this plane L becomes conjugate to the screen, i.e. the object plane of the intermediate lens should correspond to the plane L. For most microscopes, the object plane of the intermediate lens can only be placed at the level of the image plane of the objective lens (in the image mode) or at the level of the back focal plane of the objective lens (in the diffraction mode) but not on an arbitrary plane L. Thus, this technique requires a microscope permitting free control of the intermediate lens. Tanaka also proposed a variant of this technique where all the diffracted disks are simultaneously set at the exact Bragg orientation [IX.7] as shown on the example given on figure IX.17.

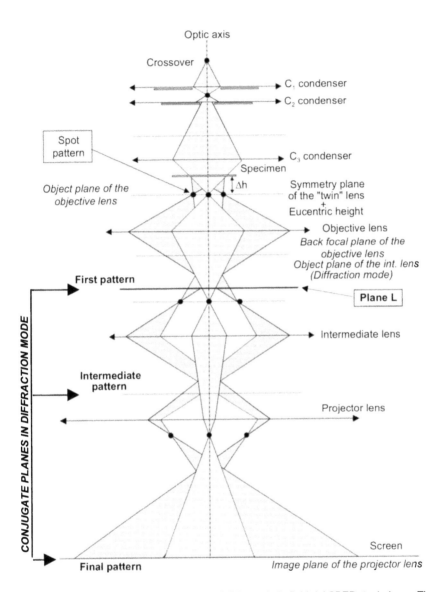

Figure IX.16 - Electron ray-paths for the bright + dark field LACBED technique. The transmitted and diffracted disks do not overlap at the level of the plane L. This plane is made conjugate to the screen by adjusting the intermediate lens so that its object plane and the plane L coincide. From Tanaka et al. [IX.4].

Figure IX.17 - Example of bright + dark field LACBED pattern. It is composed of a transmitted disk and several non-overlapping diffracted disks. All the diffracted disks are simultaneously at the exact Bragg orientation.
Pattern from a silicon specimen. Courtesy of M. Tanaka.

    This type of pattern is particularly useful for the analysis of the point group of a crystal since the latter requires the identification of symmetries present inside the bright- and dark-field disks [i.3].

IX.2 - Serial methods

IX.2.1 - The beam-rocking method (SACP)

This technique, known as the beam-rocking technique or SACP (Selected-Area Channelling Pattern), was simultaneously proposed by Eades [IX.8] and Tanaka et al.[i.6]. A parallel incident beam is rocked on a point P located on the specimen with the aid of the pre-specimen beam deflection coils (Figures VIII.11 and IX.18).

For each orientation of this incident beam, a spot pattern is obtained in the back focal plane of the objective lens. This pattern moves at the scanning rate (Figure IX.19a). It is immobilized with the post-specimen beam deflection coils, which are adjusted in such a way as to compensate exactly the scanning of the incident beam (Figure IX.19b).

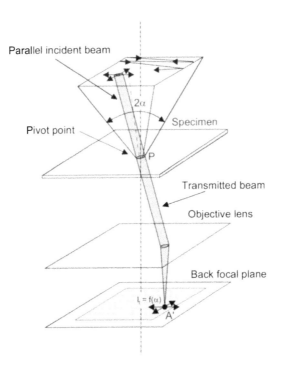

Figure IX.18 - Schematic description of the beam-rocking method.
A parallel incident beam with a small diameter rocks about a rocking point P located on the specimen. The spot pattern moves in the back focal plane at the scanning rate.
For the sake of clarity, only the transmitted beam is represented on this figure.

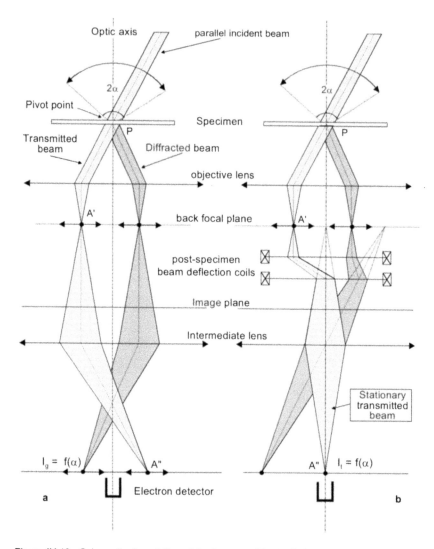

Figure IX.19 - Schematic description of the beam-rocking technique.
a - The parallel incident beam has a small diameter and rocks about a point P located on the specimen thanks to the pre-specimen beam deflection coils. The spot pattern moves at the scanning rate in the back focal plane of the objective lens.
b - The pre-specimen scanning is exactly compensated by a second post-specimen scanning. The diffraction pattern is stationary in the plane containing the detector. Note that only the transmitted beam is recorded by the detector.

The transmitted beam is then directed along the optic axis where an electron detector records the transmitted intensity I_t as a function of the orientation of the incident beam with respect to the specimen. Note that the diffracted beams are not "seen" by the detector.

The amplified signal is used to form a point-by-point image on a monitor synchronized with the scanning device of the incident beam to give a bright-field LACBED pattern (Figure IX.20). Dark-field patterns and mixed patterns can also be obtained by using central or annular electron detectors.

Figure IX.20 - Experimental beam-rocking pattern. [001] zone-axis pattern from a silicon specimen. The convergence semi-angle is about 5.5°.

The main advantage of this technique is that the transmitted and diffracted disks are not superimposed since only the transmitted beam is recorded by the detector. Large angular scans up to several degrees can be obtained with this technique. The pattern shown on figure IX.20 has a convergence semi-angle of about 5°. However, the beam-rocking technique remains relatively cumbersome to perform and is not well adapted to the analysis of defects since it is not a defocused technique.

IX.2.2 - The specimen-rocking technique

It is also possible to rock the specimen instead of the beam (Figure IX.21). Technical problems connected with the goniometer stage movements make it difficult to carry out this technique [IX.9, 10].

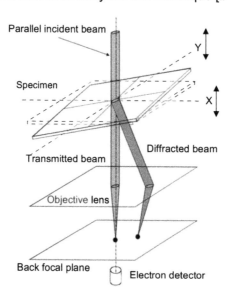

Figure IX.21 - The specimen-rocking technique. The specimen rocks around the two perpendicular X and Y axes. The detector records the transmitted intensity as a function of the orientation of the specimen.

IX.2.3 - Montage of CBED patterns

This technique, used by Rackham and Eades [IX.10], consists in obtaining the transmitted disk of a LACBED pattern by a montage of a set of CBED patterns (for which the disks do not overlap) obtained by changing the orientation of a small convergent incident beam inside a cone with a larger convergence (Figure 22a). The modification of the beam orientation is made with the pre-specimen beam deflection coils of the microscope. An example is given on figure 22b. The montage of CBED patterns is identical with the LACBED pattern on figure 22c. Note the superior quality of the LACBED pattern resulting from inelastic filtering by the selected-area aperture. Weak deficiency lines are more visible.

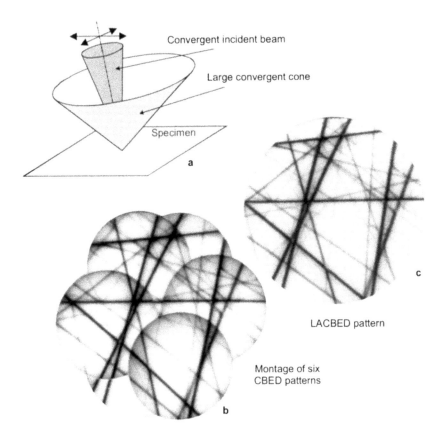

Figure IX.22 - Montage of CBED patterns.
a - Illumination conditions.
b - Experimental CBED pattern obtained by the montage of six individual CBED patterns.
c - Corresponding LACBED pattern.
Note that some weak deficiency lines are more visible on the LACBED pattern than on the montage.

IX.3 - Filtering LACBED patterns

Incident electrons can undergo elastic and inelastic interactions with the specimen and some of them lose energy and/or change direction
Four categories of inelastic scattering can be distinguished:
- thermal diffuse scattering,

- plasmon scattering,
- inner-shell loss scattering,
- Compton scattering.

Inelastic scattering has consequences both on images and diffraction patterns. It reduces the image quality, because of the chromatic aberration related to electron energy losses. On diffraction patterns, it results in the presence of a diffuse background caused by the electrons suffering a direction change. These disturbances are more significant when the specimen is thick since the probability of inelastic interactions increases with the specimen thickness.

IX.3.1 - Filtering electrons

Electrons can be removed by "angular" or by "energy" filtering.

IX.3.1.1 - Angular filtering

In this type of filtering, the electrons are filtered with respect to their direction by means of an aperture. In the image mode, the objective aperture (also called the contrast aperture) plays this role. It removes all the electrons scattered at the exit face of the specimen with an angle ρ higher than the aperture angle (Figure IX. 23a). In the diffraction mode, the selected-area aperture acts as an efficient angular filter in the LACBED technique (Figure IX.23b). It also removes electrons scattered with an angle higher than ρ. The other electron diffraction techniques do not benefit from this angular filtering.

IX.3.1.2 - Energy filtering

Electrons are filtered with magnetic or electrostatic devices, which deviate them according to their energy. Currently, three types of energy-filtering devices are available and are inserted either inside the microscope column (in-column filter) or outside it (post-column filter):
- the "Castaing-Henry" filter, which consists of a double magnetic prism coupled to an electrostatic mirror. A commercial model is available in the Zeiss 902 microscopes and only operates at accelerating voltages lower than 80 kV.
- the "omega" filter. This magnetic device is available in LEO microscopes and operates at higher voltages.
- The "GIF" (Gatan Imaging Filter) filter. This is again a magnetic device now installed in a "post-column" position. It does not affect the intrinsic performance of the microscope.

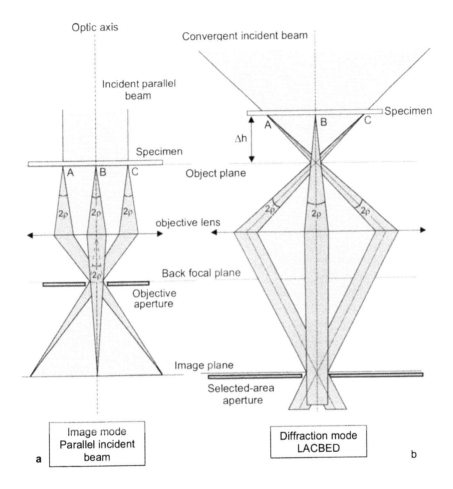

Figure IX.23 - Angular filtering of electrons.
a - In the image mode with a parallel incident beam. The objective aperture limits the aperture angle ρ of the rays at the exit face of the specimen. Inelastic electrons scattered with an angle higher than ρ are eliminated by this aperture.
b - In the LACBED method. The selected-area aperture removes the inelastic electrons scattered with an angle higher than ρ.
In order to simplify these drawings, only the ray-paths connected with the three points A, B and C at the exit face of the specimen are shown.

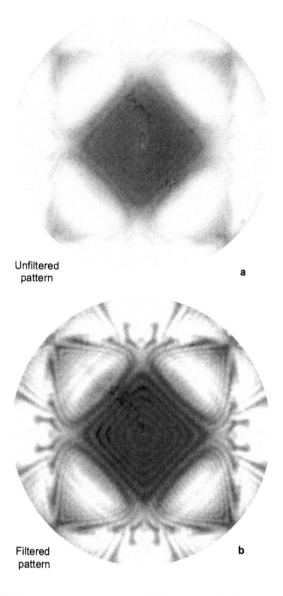

Unfiltered
pattern

a

Filtered
pattern

b

Figure IX.24 - Effect of energy filtering on LACBED patterns. [001] zone-axis pattern from a silicon specimen.
a - Unfiltered pattern.
b - Energy-filtered pattern obtained with the omega filter of the LEO 912 microscope. The improvement is striking.

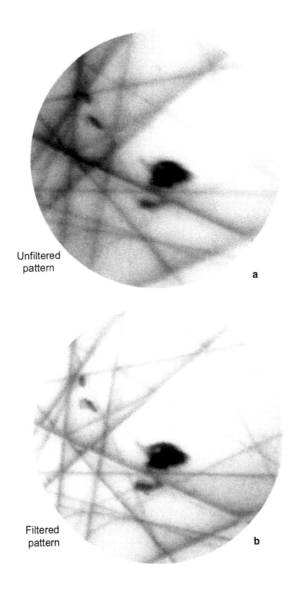

Unfiltered
pattern

a

Filtered
pattern

b

Figure IX.25 - Energy-filtered CBIM pattern.
a - Unfiltered pattern.
b - Energy-filtered pattern obtained with the omega filter of the LEO 912 microscope.

IX.3.1.2.1 - Effect of energy filtering on LACBED patterns

LACBED patterns are characterized by an excellent quality, mainly due to the strong filtering effect of the inelastic electrons by the selected-area aperture. However, this aperture eliminates only the inelastic electrons that undergo a sufficiently large change of direction. Electrons close to the optic axis are not eliminated. The effect is all the more significant since the diameter of the selected-area aperture is small and/or Δh is large (see paragraph VI.7).

The quality of LACBED patterns can be improved by removing the inelastic electrons transmitted by the selected-area aperture. The LACBED pattern on figure IX.24b, has been energy filtered using the omega filter of the LEO 912 microscope. It is a considerable improvement on the unfiltered pattern on figure IX.24a. The effect of energy filtering is important for specimens thickness than several hundreds of nanometres.

The main interest of energy filtering is to give access to the field of **quantitative electron diffraction** [IX.12 and 13]. Experimental measurements of transmitted and diffracted "elastic" intensities performed with a CCD camera or "Imaging plates" allow the determination of:
- the crystal potential (the position and the nature of the atoms in the unit cell, i.e. the structure factor),
- the Debye-Waller factor,
- the bond charge density,
at microscopic and nanoscopic scales.

Energy-filtered LACBED patterns have a unique property; energy filtering removes all inelastic electrons except for the thermally scattered electrons since their energy loss is too small (<10 meV). In fact, on LACBED patterns, these thermally electrons are also eliminated by angular filtering with the selected-area aperture. This means that the suppression of inelastic scattering is complete on energy-filtered LACBED patterns.

IX.3.1.2.2 - Effect of energy filtering on CBIM patterns

The quality of CBIM patterns is inferior to that of LACBED patterns. For the former, the objective aperture is used instead of the selected-area aperture. In order to avoid too strong a reduction of the convergence angle, the size of the objective aperture must be large, with the result that the angular filtering is poor. In this case, an additional energy filtering produces a substantial quality improvement as illustrated on figure IX.25.

CHAPTER X
Indexing LACBED patterns

X.1 - Generalities

Bright-field LACBED patterns usually contain a very large number of deficiency lines (up to several hundreds). This wealth of information is a great advantage compared to the other electron diffraction techniques. Nevertheless, it can be a drawback for pattern indexing. Moreover, a deformed specimen produces distorted LACBED patterns. The excess and deficiency lines present on the pattern are then more or less bent and the angles between them modified.

Which pieces of information are available on LACBED patterns?

- The pattern symmetry can be examined in order to identify, for example, the point group. This type of examination is generally done on zone-axis patterns. In this case, we just need to identify the zone-axis indices [uvw].

- Specific excess or deficiency Bragg lines can be identified. For example, we can observe the effect of a dislocation on a given Bragg line in order to characterize its Burgers vector. In this case, we need to identify the hkl indices of this Bragg line. If the line is located on a zone-axis pattern or near a zone axis, the identification is relatively easy. If not, diffraction patterns must be simulated.

X.2 – Computer simulation of LACBED patterns

Two categories of computer simulations are available: dynamical and kinematical simulations.

X.2.1 - Dynamical simulations

All the dynamical aspects developed in chapter VII are taken into account in the simulation of these patterns.

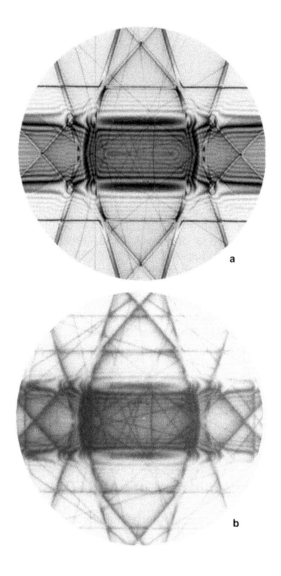

Figure X.1 - Dynamical simulation of a [113] LACBED pattern from a silicon specimen.
a - Simulated LACBED pattern obtained with the Jems software from P. Stadelmann [X.1].
The simulation was carried out for a specimen thickness of 350 nm.
b - Corresponding experimental LACBED pattern.
The agreement between the simulated and the experimental patterns is excellent.

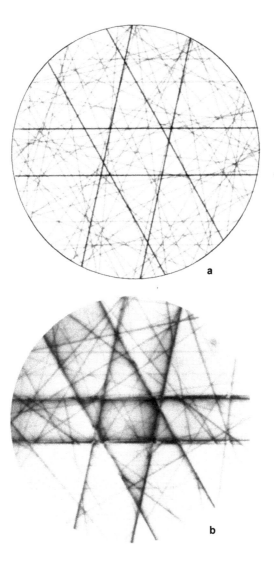

Figure X.2 - Kinematical simulation of a [345] LACBED pattern from a silicon specimen.
a - LACBED pattern simulated with the "Electron Diffraction" software [VI.3].
Dynamical lines are represented by bold lines and quasi-kinematical lines by fine lines. The simulation was carried out for a specimen thickness of 200 nm.
b - Corresponding LACBED pattern.

These simulations use Bloch waves or the multislice theories. The [113] zone-axis pattern given on figure X.1a was simulated with the Jems software developed by P. Stadelmann [X.1]. It illustrates the perfect agreement between the simulated and experimental patterns.

Since dynamical phenomena depend greatly on the specimen thickness, dynamical simulations must be carried out for well-defined thickness values.

The advantages of this type of simulation are obvious. They reproduce perfectly all the dynamical effects. Unfortunately, these simulations require rather long calculations (from a few minutes to several hours) and powerful computers and are hence not well adapted to the fast indexing of Bragg lines.

X.2.2 - Kinematical simulations

The general principle used to obtain kinematical simulations consists in determining the lattice planes of the specimen at the Bragg orientation and calculating the positions of their corresponding excess and deficiency lines in the back focal plane of the objective lens. The quality of the simulated patterns can be improved by calculating the kinematical intensity of the Bragg lines, i.e. the squared modulus of the structure factors $|F_{hkl}|^2$. The useless weak Bragg lines can then be removed from the drawing. The extinction distance ξ_g can also be calculated in order to distinguish between the quasi-kinematical and the dynamical lines described in chapter VII.5. We recall that dynamical lines consist of a set of fringes and occur when $t > \xi_g / 3$. They can be represented by thick lines on the simulations. Quasi-kinematical lines, however, only display a single minimum and occur when $t < \xi_g / 3$. They can be represented by sharp lines. Thanks to these improvements, the kinematical simulated patterns bear a good resemblance to the experimental patterns.

This type of simulation requires very short calculation times (a few seconds). An example of a kinematical simulation produced with the "Electron Diffraction" software [VI.3] is given on figure X.2a.

X.3 - Conventions

The signs of the (hkl) and [uvw] indices of the Bragg lines and zone axes are governed by conventions. In this book, the following conventions are used (Figure X.3):
- the electrons propagate downwards,

- the specimen orientation **S** = [u$_S$v$_S$w$_S$] is in the opposite direction, i.e. upwards,
- photographic plates are printed in such a way that they reproduce the images and the diffraction patterns as they appear on the microscope screen.

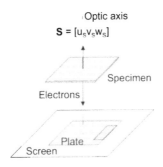

Figure X.3 - Conventions used in this book.
- The electrons move downwards.
- The specimen orientation **S** = [u$_S$v$_S$w$_S$] is directed upwards.
- The plates are printed so that the images and diffraction patterns appear as they are observed on the screen of the microscope.

X.4 - Indexing LACBED patterns

The indexing methods can be separated into two categories according to whether the pattern is a zone-axis pattern or not.

X.4.1 - Indexing [uvw] zone-axis LACBED patterns

The pattern on figure X.4a is a good example of an experimental zone-axis LACBED pattern (it is also used in chapter XI to characterize dislocations). This pattern is strongly distorted and the Bragg lines are very broad, especially in the thinnest areas of the specimen. It is nearly impossible to index it directly.

The first step consists in obtaining an undistorted pattern. If possible, another pattern with the same orientation is produced from a thicker and less distorted specimen area. The pattern on figure X.2b was produced in this way. Another solution is to obtain the corresponding Kossel pattern (Figure X.4b). The latter has a poorer quality than the LACBED pattern, but it has the major advantage of not being distorted Consequently, the angles and the distances between the lines are in agreement with the theoretical patterns.

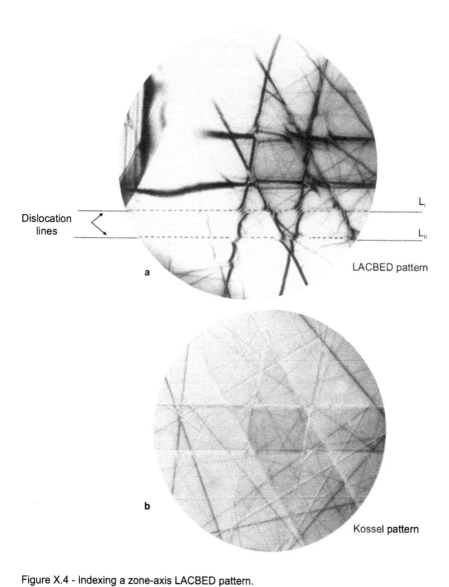

Dislocation
lines

L$_I$

L$_{II}$

a

LACBED pattern

b

Kossel pattern

Figure X.4 - Indexing a zone-axis LACBED pattern.
a - Experimental LACBED pattern. This pattern is very strongly distorted owing to deformation and thickness variations in the illuminated area of the specimen. The traces of two dislocation lines L$_I$ and L$_{II}$ are visible. They are characterized in paragraph XI.1.2.1.1.
b - Corresponding Kossel pattern. The contrast is poor, but the pattern is undistorted.
Silicon specimen. Courtesy of J.P. Michel.

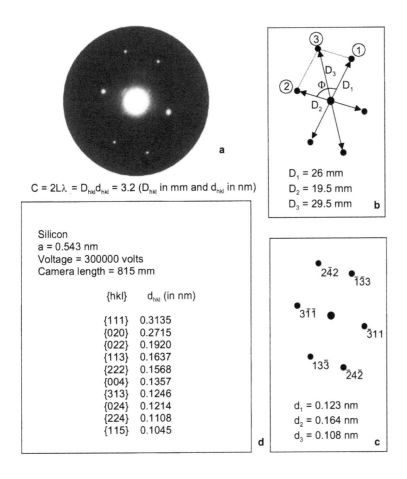

$C = 2L\lambda = D_{hkl}d_{hkl} = 3.2$ (D_{hkl} in mm and d_{hkl} in nm)

$D_1 = 26$ mm
$D_2 = 19.5$ mm
$D_3 = 29.5$ mm

b

Silicon
a = 0.543 nm
Voltage = 300000 volts
Camera length = 815 mm

{hkl}	d_{hkl} (in nm)
{111}	0.3135
{020}	0.2715
{022}	0.1920
{113}	0.1637
{222}	0.1568
{004}	0.1357
{313}	0.1246
{024}	0.1214
{224}	0.1108
{115}	0.1045

$2\bar{4}2$ $\bar{1}\bar{3}3$

$3\bar{1}\bar{1}$

311

$13\bar{3}$ $\bar{2}4\bar{2}$

$d_1 = 0.123$ nm
$d_2 = 0.164$ nm
$d_3 = 0.108$ nm

c

d

Figure X.5 - Indexing the microdiffraction pattern corresponding to the LACBED pattern on figure X.4.
a - Experimental microdiffraction pattern obtained for L = 815 mm and V = 300 kV. The diffraction constant is: $C = 2L\lambda = D_{hkl}\,d_{hkl} = 3.2$ (D_{hkl} in mm and d_{hkl} in nm).
b - Selection and measurement of the distances D_1, D_2 and D_3.
c - Indexed diffraction pattern. The cross-product $(\bar{1}\bar{3}3) \times (3\bar{1}\bar{1})$ of the 1 and 2 reflections gives the [345] zone axis.
d - d-spacing table for a silicon specimen.
The simulation and the d-spacings table were obtained with the "Electron Diffraction" software [VI.3].

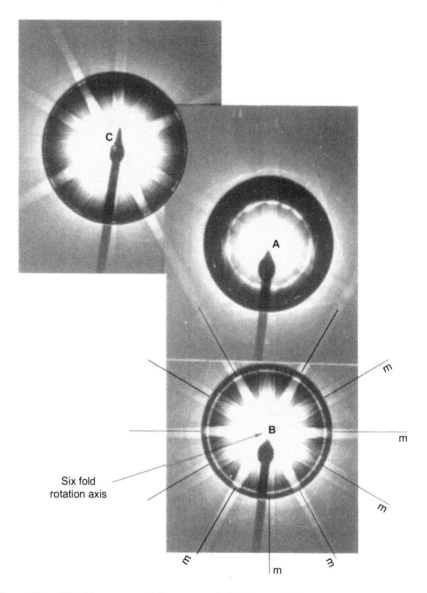

Figure X.6 - Kikuchi map around the zone axis **A** = [$u_A v_A w_A$]. The zone axis **B** = [$u_B v_B w_B$] displays a 6mm "projected" symmetry typical of a <111> zone axis for a cubic crystal.
The 6mm symmetry is characterized by a 6-fold rotation axis (60° rotation) and 6 mirrors m (symmetry lines) located at 30° from one another.

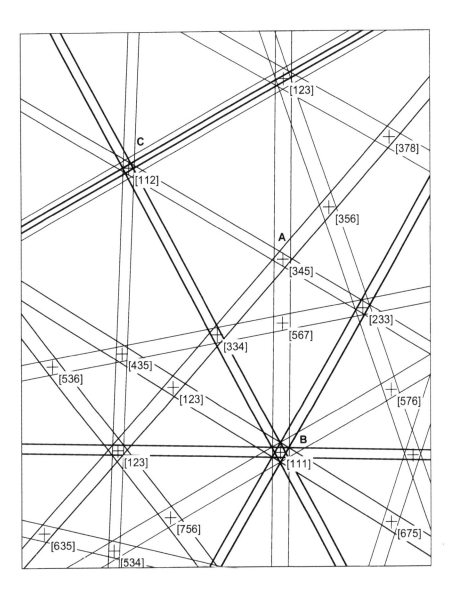

Figure X.7 - Simulated Kikuchi map around the [111] zone axis corresponding to the experimental patterns on figure X.6. The zone axis under examination **A** is the [345] zone axis. Simulation carried out with the "Electron Diffraction" software [VI.3].

X.4.1.1 - Identification of the [uvw] zone axis

The [uvw] zone axis can be identified from spot patterns or from Kikuchi patterns.

X.4.1.1.1 - Identification of the [uvw] zone axis from spot patterns

The method consists in producing a selected-area or a microdiffraction spot pattern with a parallel or almost parallel incident beam and indexing this pattern according to the following steps (Figure X.5a).

- Three reflections (labelled 1, 2 and 3) close to the origin and disposed as shown on figure X.5b, are selected.
- The distances D_1, D_2 and D_3 between these reflections and their symmetrical reflections are measured (Figure X.5b).
- Using the relation $D_{hkl} \, d_{hkl} = 2L\lambda = C$ (III.5), d_1, d_2 and d_3 spacings are assigned to the 3 reflections (Figure X.5c).
- From a table of calculated d-spacings (Figure X.5d), hkl indices are assigned to each reflection ensuring that:
$h_3 = h_1 + h_2$
$k_3 = k_1 + k_2$
$l_3 = l_1 + l_2$
and that the angle Φ between the reflections 1 and 2 agrees with the calculated angle. The indices of the other reflections are easily obtained by adding the first three indices as indicated on figure X.5c.
- The [uvw] zone axis is identified from the $(h_1 k_1 l_1) \times (h_2 k_2 l_2)$ cross-product between the reflections 1 and 2:
$u = k_1 l_2 - l_1 k_2$
$v = - (h_1 l_2 - l_1 h_2)$
$w = h_1 k_2 - k_1 h_2$
In the present case, we obtain [345].

Note
To be in agreement with the conventions given on figure X.3 and obtain a [uvw] zone axis directed upwards, the $h_1 k_1 l_1$ and $h_2 k_2 l_2$ reflections must be chosen as shown on figure X.5b.

X.4.1.1.2 - Identification of the [uvw] zone axis from Kikuchi patterns.

We can also tilt the specimen around the zone axis $\mathbf{A} = [u_A v_A w_A]$ until a highly symmetric Kikuchi pattern is obtained. In the present case, we obtain the Kikuchi map on figure X.6. The zone axis $\mathbf{B} = [u_B v_B w_B]$

displays a 6mm* "projected" symmetry (a sixfold rotation axis and 6 mirrors at 30° from one another), typical of a <111> zone axis for a cubic crystal. Thanks to the corresponding simulation of figure X.7, the [345] zone axis is easily identified.

(*) The true symmetry of a <111> zone axis pattern for a cubic crystal is 3m and not 6mm (a threefold rotation axis and 3 mirrors at 120° from one another). In the present case, the B pattern displays a 6mm symmetry because we observe only its 2D "projected" symmetry and not its 3D symmetry. Details on 2D and 3D symmetries are given in reference [i.3].

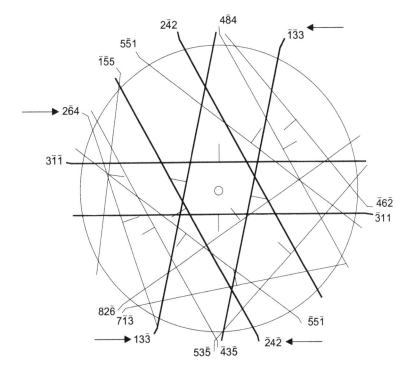

Figure X.8 - Simulation of a [345] LACBED pattern from a silicon specimen.
In order to simplify this drawing, only the strongest Bragg lines are displayed. The small lines perpendicular to the Bragg lines indicate the positive direction of the deviation parameter s. The lines indicated with an arrow are used in paragraph XI.1.2.1.1 to identify the Burgers vector of dislocations.
Simulation carried out with the "Electron Diffraction" software [VI.3].

A zone-axis LACBED pattern can then be simulated and each of its Bragg lines indexed (Figure X.8). This is the case for the distorted LACBED pattern on figure X.3a.

X.4.2 - Indexing non-zone-axis LACBED patterns

This second example concerns LACBED patterns obtained from a garnet specimen (Figure X.9a). Garnet has a cubic structure with a body-centred Bravais lattice. Since its lattice parameter a = 1.153 nm is large, a great number of Bragg lines is observed on the experimental patterns.

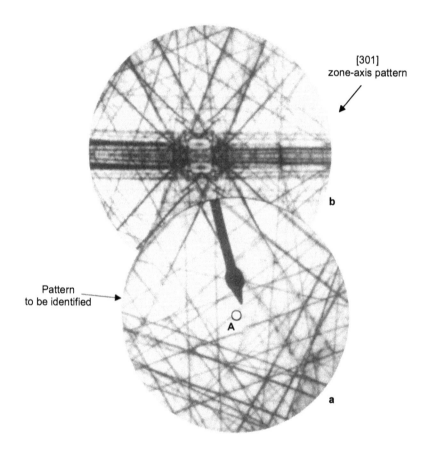

[301]
zone-axis pattern

b

Pattern
to be identified

A

a

Figure X.9 - Identification of a non-zone-axis LACBED pattern.
a - Experimental pattern.
b - Closest zone-axis pattern: [301] zone axis.
LACBED patterns from a garnet specimen. Courtesy of P. Cordier.

Here, we are attempting to identify a non-zone-axis LACBED pattern. A solution consists in tilting the specimen in order to find the closest [uvw] zone axis and to interpret it with the above methods.

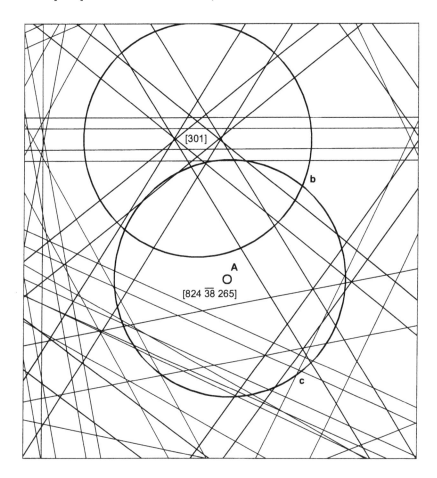

Figure X.10 - Simulation of a [301] LACBED pattern from a garnet specimen. A 10° convergence semi-angle is used in order to cover a wide angular field. The patterns a and b on Figure X.9 are indicated by circles. The point A on the simulation corresponds to the [824 $\overline{38}$ 265] direction.
Simulation carried out with the "Electron diffraction" software [VI.3].

In the present case, the [301] zone axis is identified (Figure X.9b). A LACBED simulation around this zone axis is then made with a wide convergence angle in order to cover the angular field of the experimental patterns (Figure X.10). The identification is simply carried out by comparing theoretical and experimental patterns.

X.4.2.1 - Identification of the crystal orientation S = [u_Sv_Sw_S]

In order to identify the specimen orientation $S = [u_Sv_Sw_S]$ with respect to the incident electron beam, we can use the following property. Each point of a LACBED pattern, for example the point A on figure X.11, comes from an incident ray having a well-defined orientation inside the convergent incident beam. A lattice direction $[u_Av_Aw_A]$ is directed along this beam. In the same way, the specimen orientation S corresponds to the centre of the pattern, i.e. to the point F. Therefore, we can associate a crystal direction [uvw] to any point of the LACBED pattern. The problem is to identify these lattice directions from LACBED patterns. This operation can be performed with simulated patterns obtained with the "Electron Diffraction" software [VI.3] using a method based on the analysis of the crystal orientation from Kikuchi lines.

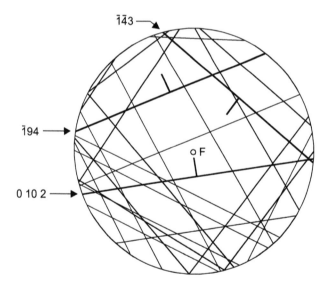

Figure X.12 - Simulation of the experimental pattern shown on Figure X.8.

The centre of the pattern (point F) corresponds to the [824 $\overline{38}$ 265] lattice direction.

The bold lines marked with an arrow are used for identifying the Burgers vector of a dislocation in paragraph XI.2.2.1.2.

The small segments indicate the positive direction of the deviation parameter s.

In order to simplify this drawing, only the strongest Bragg Lines are drawn.

Simulation carried out with the "Electron diffraction" software [VI.3].

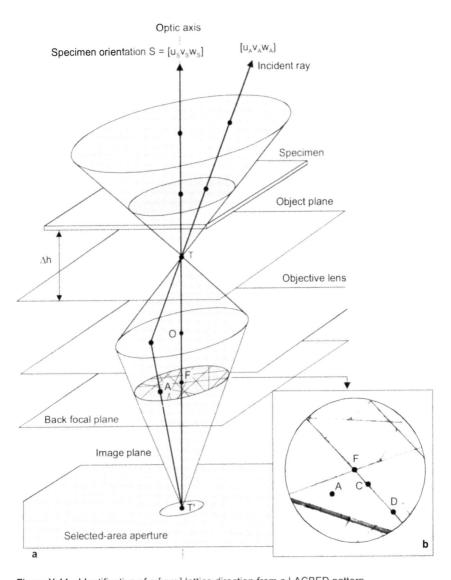

Figure X.11 - Identification of a [uvw] lattice direction from a LACBED pattern.
a - Any point of the pattern, for example the point A, corresponds to an incident ray directed along a $[u_A v_A w_A]$ lattice direction. The specimen orientation $\mathbf{S} = [u_S v_S w_S]$ corresponds to the centre of the pattern F. These directions can be identified from simulated patterns.
b - Central area of an experimental LACBED pattern. Note that the directions corresponding to the intersections of two kinematical lines (for example the points C or D) are very easy to identify from simulated patterns and give the most accurate results.

For example, the [824 $\overline{38}$ 265] lattice direction is identified for the point A on the experimental pattern of figure X.9. Once this direction is determined, the full indexing of the LACBED pattern can be performed as shown on figure X.12.

Note

The [824 $\overline{38}$ 265] lattice direction corresponds to a vector of the direct lattice

$u = 824a - 38b + 265c$ *joining the lattice origin 0 to the node* 824 $\overline{38}$ 265. *Since this node is*

located far from the origin, the [824 $\overline{38}$ 265] direction is accurately defined.

To improve the accuracy in the determination of these lattice directions it is advisable to consider the intersections of weak quasi-kinematical lines (for example the points C, D and E on figure X.13b). These particular points are very easy to localize accurately on simulated patterns. In addition, they are not affected by specimen distortion.

X.5 - Trace analyses from LACBED patterns

LACBED patterns have the unique property of giving information on both the reciprocal and the direct spaces. The shadow image of the illuminated area of the specimen is superimposed on the diffraction pattern. This property can be used to perform trace analyses directly from LACBED patterns in order to identify [uvw] directions and (hkl) planes.

The main principle used for trace analysis from LACBED patterns consists in observing Bragg lines (in the reciprocal space) parallel to the trace of the studied direction or plane (in the direct space). Alternatively, the identification of lattice directions connected with points situated on the trace can be used when parallel Bragg lines are difficult to find.

.

We will assume here that the LACBED patterns are obtained with $\Delta h < 0$ (the specimen is located below the object plane). This is a favourable situation since there is no rotation between the diffraction pattern and the shadow image. If the specimen is above the object plane ($\Delta h > 0$), then a 180° rotation must be taken into account in the analysis. We will also assume that the axis of the convergent incident beam is directed along the optic axis of the microscope. This is the situation encountered when the microscope is correctly aligned.

What information can be obtained from the Bragg lines present on a LACBED pattern?

We consider a specimen orientation $S_1 = [u_{S1}v_{S1}w_{S1}]$ so that a set of lattice planes $(h_1k_1l_1)$ is parallel to the optic axis of the microscope (Figure X.13).

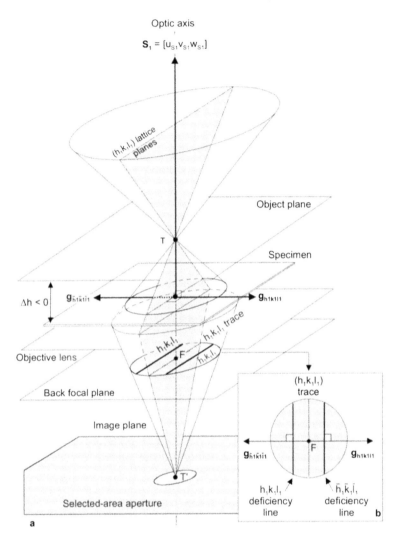

Figure X.13 - Connection between a set of vertical lattice planes $(h_1k_1l_1)$ and the LACBED pattern located in the back focal plane of the objective lens.

a - The two reciprocal vectors $g_{h_1k_1l_1}$ and $g_{\bar{h}_1\bar{k}_1\bar{l}_1}$ associated with the $(h_1k_1l_1)$ lattice planes are parallel to the back focal plane and perpendicular to the pair of $h_1k_1l_1/\bar{h}_1\bar{k}_1\bar{l}_1$ Bragg lines. The specimen is located below the object plane so that there is no rotation between the illuminated area and its shadow image.

b - Relative arrangement of the pair of $h_1k_1l_1/\bar{h}_1\bar{k}_1\bar{l}_1$ Bragg lines, the trace of the $(h_1k_1l_1)$ lattice planes and the reciprocal lattice vectors $g_{h_1k_1l_1}$ and $g_{\bar{h}_1\bar{k}_1\bar{l}_1}$.

This set $(h_1k_1l_1)$ is vertical in the microscope and it can be characterized by its two reciprocal lattice vectors, g_{h1k1l1} on one side of the set and $g_{\bar{h}1\bar{k}1\bar{l}1}$ on the other side. By definition, these two vectors are perpendicular to the $(h_1k_1l_1)$ lattice planes and their modulus is: $g_{h1k1l1} = 1/d_{h1k1l1}$, where d_{h1k1l1} is the interplanar distance. (X.1)

In the present case, g_{h1k1l1} and S_1 are perpendicular and so the dot product $g_{h1k1l1} \cdot S_1$ vanishes, i.e:

$$g_{h1k1l1} \cdot S_1 = (h_1a^* + k_1b^* + l_1c^*) \cdot (u_{S1}a + v_{S1}b + w_{S1}c) = h_1u_{S1} + k_1v_{S1} + l_1w_{S1} = 0$$

We can also consider that S_1 is a zone axis for the $(h_1k_1l_1)$ lattice planes and therefore apply the zone condition: $h_1u_{S1} + k_1v_{S1} + l_1w_{S1} = 0$.

As explained in the previous chapters, the bright-field LACBED pattern, located in the back focal plane of the objective lens, displays a pair of $h_1k_1l_1/\bar{h}_1\bar{k}_1\bar{l}_1$ deficiency Bragg lines. The two lines are parallel to the trace of the $(h_1k_1l_1)$ lattice planes, which in this particular case crosses the back focal point F. As shown on figure X.13b, the two reciprocal vectors g_{h1k1l1} and $g_{\bar{h}1\bar{k}1\bar{l}1}$ are parallel to the back focal plane and perpendicular to the Bragg lines. This means that these vectors – and consequently the $(h_1k_1l_1)$ lattice planes – can be identified provided the indices of the Bragg lines are known. In fact, this analysis remains valid for any Bragg lines on a LACBED pattern for the following reason: all the Bragg lines present on a LACBED pattern come from vertical or nearly vertical lattice planes. As a first approximation, we can consider that all the g_{hkl} vectors involved are located in the back focal plane of the objective lens.

What information can be obtained from the shadow image?
The trace of the lines or planes can be obtained from the shadow image. With the experimental conditions given above ($\Delta h < 0$), this trace is not rotated with respect to the Bragg lines.

X.5.1 - Identification of lines from LACBED patterns

The identification of lines is particularly useful for the analysis of dislocations.
A line L is characterized by a vector u whose polarity is arbitrarily chosen.

$$u = [u_uv_uw_u]$$

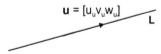

Figure 14. Characterization of a line L with a vector u whose polarity is arbitrarily chosen.

This vector is a lattice direction:

$$\mathbf{u} = [u_u v_u w_u] = u_u \mathbf{a} + v_u \mathbf{b} + w_u \mathbf{c} \qquad (X.3)$$

The identification of a lattice direction $\mathbf{u}$ consists in finding two lattice planes $(h_1 k_1 l_1)$ and $(h_2 k_2 l_2)$ containing this direction. As shown on figure X.15, the two corresponding reciprocal vectors $\mathbf{g}_{h1k1l1}$ and $\mathbf{g}_{h2k2l2}$ are perpendicular to $\mathbf{u}$ and the dot products $\mathbf{g}_{h1k1l1}.\mathbf{u}$ and $\mathbf{g}_{h2k2l2}.\mathbf{u}$ are equal to zero.

We obtain,

$$\mathbf{g}_{h1k1l1}.\mathbf{u} = (h_1 \mathbf{a}^* + k_1 \mathbf{b}^* + l_1 \mathbf{c}^*).(u_u \mathbf{a} + v_u \mathbf{b} + w_u \mathbf{c}) = h_1 u_u + k_1 v_u + l_1 w_u = 0 \quad (X.4)$$

and

$$\mathbf{g}_{h2k2l2}.\mathbf{u} = (h_2 \mathbf{a}^* + k_2 \mathbf{b}^* + l_2 \mathbf{c}^*).(u_u \mathbf{a} + v_u \mathbf{b} + w_u \mathbf{c}) = h_2 u_u + k_2 v_u + l_2 w_u = 0 \quad (X.5)$$

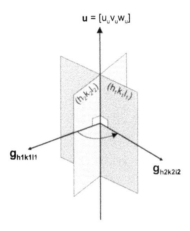

Figure X.15 - Characterization of a lattice direction $\mathbf{u} = [u_u v_u w_u]$ from the identification of two sets of lattice planes $(h_1 k_1 l_1)$ and $(h_2 k_2 l_2)$ containing this direction. The reciprocal lattice vectors $\mathbf{g}_{h1k1l1}$ and $\mathbf{g}_{h2k2l2}$ are perpendicular to the direction $\mathbf{u}$.
The cross-product $\mathbf{g}_{h1k1l1} \times \mathbf{g}_{h2k2l2}$ allows $\mathbf{u}$ to be identified.
Note that $\mathbf{g}_{h1k1l1}$ and $\mathbf{g}_{h2k2l2}$ should be arranged as shown on the figure so that the cross-product gives $\mathbf{u}$ and not $-\mathbf{u}$.

The indices u_u, v_u and w_u are obtained by solving the following system of equations:

$h_1 u_u + k_1 v_u + l_1 w_u = 0$
$h_2 u_u + k_2 v_u + l_2 w_u = 0$

or from the cross-product:

$\mathbf{g}_{h1k1l1} \times \mathbf{g}_{h2k2l2}$ giving:

$u_u = k_1 l_2 - l_1 k_2$
$v_u = - (h_1 l_2 - l_1 h_2)$
$w_u = h_1 k_2 - k_1 h_2$ (to within a constant)

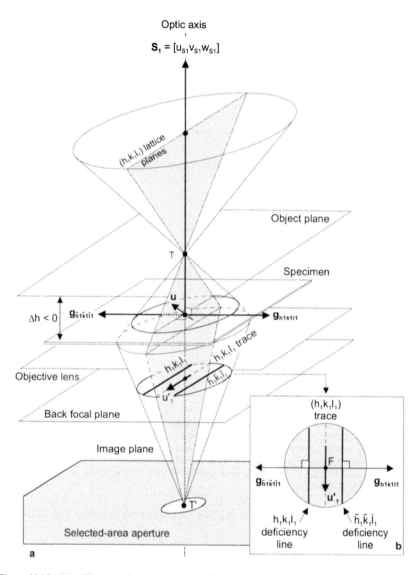

Figure X.16 - Identification of a line **u** from LACBED patterns.

a - The trace **u'₁** of the line **u** is parallel to the pair of h₁k₁l₁/h̄₁k̄₁l̄₁ Bragg lines. This allows the (h₁k₁l₁) lattice plane containing the line **u** to be identified.

b - Relative arrangement of the pair of h₁k₁l₁/h̄₁k̄₁l̄₁ Bragg lines, the **u'₁** trace of the **u** line and the reciprocal lattice vectors g_{h1k1l1} and g_{h̄1k̄1l̄1}.

Several experimental methods can be proposed to identify the two lattice planes required for the analysis. We describe two of them for the general case of a line **u** tilted with respect to the optic axis.

X.5.1.1 - First method (method L₁)

The specimen is tilted to a first orientation S_1 so that the trace u'_1 of the line studied is parallel to a pair of Bragg lines $h_1k_1l_1/\overline{h_1k_1l_1}$. According to figure X.16, this means that the line **u** is contained in the $(h_1k_1l_1)$ lattice planes.

The specimen is then tilted to a second orientation S_2 yielding another pair of Bragg lines $h_2k_2l_2/\overline{h_2k_2l_2}$ parallel to the trace of the line u'_2. The second lattice plane $(h_2k_2l_2)$ required for the characterization of **u** is then identified.

It can sometimes be difficult or even impossible to find Bragg lines parallel to the line trace. In this case, the method L_2 described in the following paragraph is a good alternative since it can be performed for any specimen orientation.

X.5.1.1 - Second method (method L₂)

We pointed out in section X.4.2.1, that a lattice direction [uvw] can be associated with any point of a LACBED pattern. Let us consider two points A'_1 and B'_1 situated on the trace u'_1 of the line **u** under examination for a specimen oriented along S_1. The two corresponding lattice directions $u_{A1} = [u_{A1}v_{A1}w_{A1}]$ and $u_{B1} = [u_{B1}v_{B1}w_{B1}]$ can be identified as indicated in section X.4.2.1. In agreement with figure X.17, their cross-product $u_{A1} \times u_{B1}$ gives a reciprocal vector g_{h1k1l1} perpendicular both to u_{A1} and u_{B1} and to the set of lattice planes $(h_1k_1l_1)$. The Miller indices of the $(h_1k_1l_1)$ planes containing the line **u** are then directly identified.

The specimen is then tilted to another orientation S_2 and a second experiment is performed giving the directions u_{A2} and u_{B2} and a second set of lattice planes $(h_2k_2l_2)$ also containing the line **u**.

X.5.1.3 - Derivation of the indices u_u, v_u and w_u of the line

The two previous methods allow the identification of two $(h_1k_1l_1)$ and $(h_2k_2l_2)$ lattice planes containing the line **u**. The indices u_u, v_u and w_u of this line are obtained by solving the equations:

$h_1u_u + k_1v_u + l_1w_u = 0$
$h_2u_u + k_2v_u + l_2w_u = 0$
or from the cross product $g_{h1k1l1} \times g_{h2k2l2}$

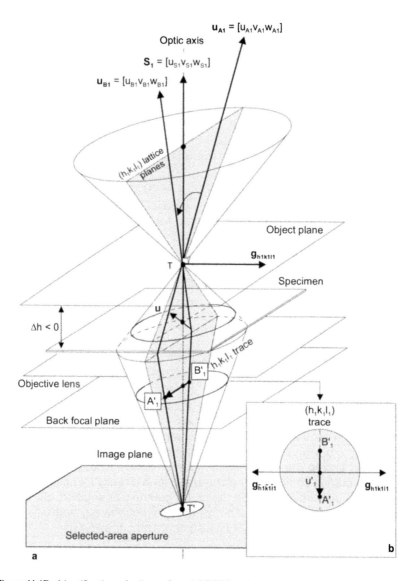

Figure X.17 - Identification of a line **u** from LACBED patterns.

a - The cross-product $\mathbf{u}_{A1} \times \mathbf{u}_{B1}$ of the two lattice directions $\mathbf{u}_{A1}$ and $\mathbf{u}_{B1}$ corresponding to the points A'$_1$ and B'$_1$ situated on the trace $\mathbf{u}'_1$ gives a reciprocal vector $\mathbf{g}_{h1k1l1}$ perpendicular to both $\mathbf{u}_{A1}$ and $\mathbf{u}_{B1}$ and to the set of lattice planes $(h_1k_1l_1)$. This allows the $(h_1k_1l_1)$ lattice planes containing the line **u** to be identified.

b - Relative arrangement of the points A'$_1$ and B'$_1$ located on the $\mathbf{u}'_1$ trace of the line **u** with the reciprocal lattice vectors $\mathbf{g}_{h1k1l1}$ and $\mathbf{g}_{\bar{h}1\bar{k}1\bar{l}1}$.

286

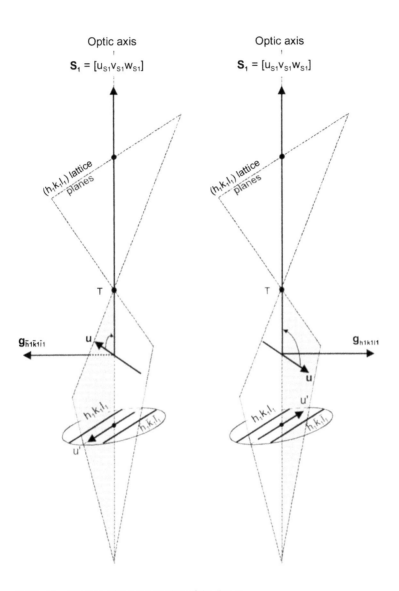

Figure X.18 - Identification of the orientation of the line **u**.

a - First configuration. The line **u** is located in the left shaded area of the $(h_1k_1l_1)$ lattice planes. The trace **u'** points forwards and the cross product **u** x **S₁** gives $\mathbf{g_{\overline{h_1}\overline{k_1}\overline{l_1}}}$.

b - Second configuration. The line **u** is located in the right shaded area of the $(h_1k_1l_1)$ lattice planes. The trace **u'** points backwards and the cross-product **u** x **S₁** gives $\mathbf{g_{h_1k_1l_1}}$.

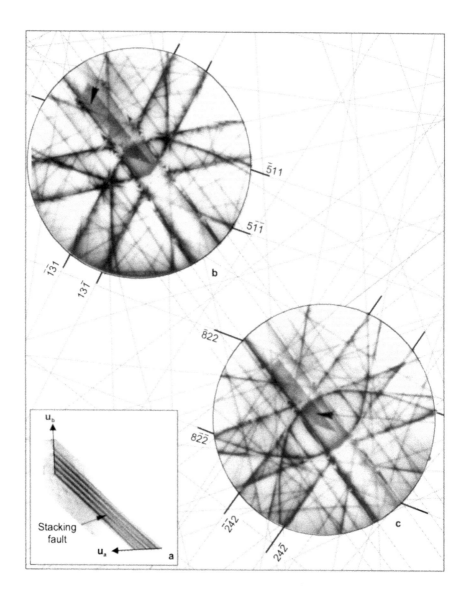

Figure X.19 - Identification of two partial dislocation lines u_a and u_B bordering a stacking fault.
a - Micrograph of the stacking fault bordered by the two partial dislocations obtained for the [00̄1] specimen orientation.
b and c - [114] and [113] LACBED patterns.
Silicon specimen. Courtesy of P.H. Albarède.

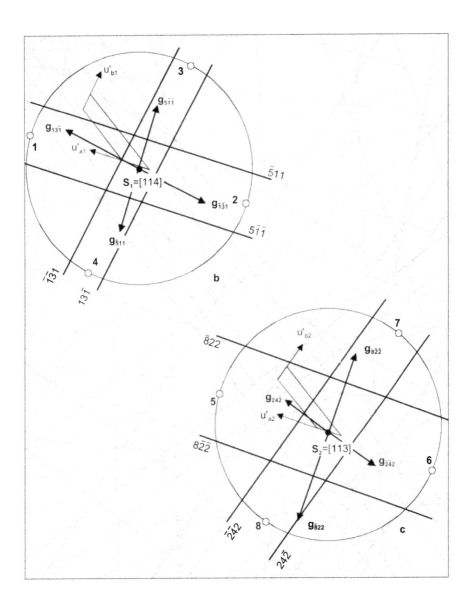

Figure X.19 - Continuation.

The traces of the u_a and u_b lines are parallel to the $5\bar{1}\bar{1}/\bar{5}11$, $\bar{8}22/8\bar{2}\bar{2}$, $13\bar{1}/\bar{1}3\bar{1}$ and $24\bar{2}/\bar{2}\bar{4}2$ pairs of Bragg lines respectively. Note that the dislocation lines and the Bragg lines are not always exactly parallel because of specimen distortion.
The indices of the corresponding reciprocal vectors are also indicated.

Nevertheless, these methods do not give the sign of **u**. To obtain it we can consider the cross-product of the vector **u** with the specimen orientation $\mathbf{S_1}$ (or $\mathbf{S_2}$). This yields a reciprocal lattice vector $\mathbf{g_{h1k1l1}}$ perpendicular both to **u** and $\mathbf{S_1}$ and directed so that the vectors **u**, $\mathbf{S_1}$, $\mathbf{g_{h1k1l1}}$ form a right-handed set. According to the two configurations a and b displayed on figure X.18 where the vector **u** is either contained in the left (a) or in the right (b) area of the $(h_1k_1l_1)$ planes (shaded area on figure X.17), this cross product is typical of the orientation of **u** since it either gives $\mathbf{g_{h1k1l1}}$ or $\mathbf{g_{\bar{h}1\bar{k}1\bar{l}1}}$.

X.5.1.4 - Application of the method L_1

The example given on figures X.19 relates to two partial dislocation lines $\mathbf{u_a}$ and $\mathbf{u_b}$ bordering a stacking fault (Figure X.19a). To begin with, the polarity of the dislocation lines $\mathbf{u_a}$ and $\mathbf{u_b}$ are chosen arbitrarily.

The observation of the $\mathbf{u_a}$ line on the $\mathbf{S_1}$ = [114] and $\mathbf{S_2}$ = [113] zone-axis LACBED patterns shows that its traces $\mathbf{u'_{a1}}$ and $\mathbf{u'_{a2}}$ are parallel to pairs of the Bragg lines $5\bar{1}\bar{1}/\bar{5}11$ and $\bar{8}22/8\bar{2}\bar{2}$ respectively. Solution of the equations:

$5u_a - v_a - w_a = 0$
$8u_a - 2v_a - 2w_a = 0$

or the cross-product $\mathbf{g_{5\bar{1}\bar{1}}}$ x $\mathbf{g_{8\bar{2}\bar{2}}}$

gives $\mathbf{u_a}$ = $[0\bar{1}1]$ or $[01\bar{1}]$ (to within a constant).

In the same way, the $\mathbf{u_b}$ traces are parallel to the pairs $13\bar{1}/\bar{1}3\bar{1}$ on the $\mathbf{S_1}$ pattern and $24\bar{2}/\bar{2}4\bar{2}$ on the $\mathbf{S_2}$ pattern giving the equations:
$u_b + 3v_b - w_b = 0$
$2u_b + 4v_b - 2w_b = 0$
whose solution is $\mathbf{u_b}$ = [101] or $[\bar{1}0\bar{1}]$ (to within a constant).

X.5.1.5 - Application of the method L_2

As an example of this method, we have selected four pairs of points (points 1 to 8) situated on the traces of the $\mathbf{u_a}$ and $\mathbf{u_b}$ lines on the patterns on figure X.19. The corresponding directions and cross-products are listed in table X.1.
They also lead to:
$\mathbf{u_a} = [01\bar{1}]$ or $[0\bar{1}1]$ and $\mathbf{u_b}$ = [101] or $[\bar{1}0\bar{1}]$.

To obtain the sign of $\mathbf{u_a}$ and $\mathbf{u_b}$, we consider the cross-product $\mathbf{u} \times \mathbf{S}$ (with $\mathbf{S} = \mathbf{S_1}$ or $\mathbf{S_2}$)

Line	Points	Direction	Cross products	Cross products
u'_{a1}	1	[242 213 1000]	$(\bar{5}11)$ or $(5\bar{1}1)$	
	2	[258 290 1000]		$[0\bar{1}1]$ or $[01\bar{1}]$
u'_{a2}	5	[323 291 1000]	$(\bar{4}11)$ or $(4\bar{1}\bar{1})$	
	6	[345 375 1000]		
u'_{b2}	3	[213 262 1000]	$(13\bar{1})$ or $(\bar{1}3\bar{1})$	
	4	[286 238 1000]		$[101]$ or $[\bar{1}0\bar{1}][$
u'_{b2}	7	[298 353 1000]	$(12\bar{1})$ or $(\bar{1}\bar{2}1)$	
	8	[369 317 1000]		

Table X.1 - Identification of the [uvw] directions of the two partial dislocation lines $\mathbf{u_a}$ and $\mathbf{u_B}$ with the L_2 method. The cross products of the directions corresponding to the 8 points shown on figure X.18d permit identification of the lattice planes containing the lines $\mathbf{u_a}$ and $\mathbf{u_B}$. The cross-product of these lattice planes gives the directions of the lines $\mathbf{u_a}$ and $\mathbf{u_B}$.

According to figure X.18, the cross-products $\mathbf{u_a} \times \mathbf{S_1}$ and $\mathbf{u_a} \times \mathbf{S_2}$ should give $8\bar{2}\bar{2}$ and $5\bar{1}\bar{1}$ respectively. This is the case with $\mathbf{u_a} = [0\bar{1}\bar{1}]$. In the same way the cross products $\mathbf{u_b} \times \mathbf{S_1}$ and $\mathbf{u_b} \times \mathbf{S_2}$ should give $2\bar{4}2$ and $1\bar{3}1$ respectively. The direction $\mathbf{u_b} = [101]$ is in agreement with these values.

X.5.2 - Identification of planes from LACBED patterns

The identification of planes concerns stacking faults, twins, grain and coincidence boundaries. A plane P is characterized by its (hkl) indices or by its normal $\mathbf{n}$, the polarity of which is chosen arbitrarily (Figure X.20). This normal $\mathbf{n}$ is a reciprocal lattice vector $\mathbf{g_{hkl}}$ with:
$$\mathbf{n} = \mathbf{g_{hkl}} = h\mathbf{a^*} + k\mathbf{b^*} + l\mathbf{c^*} \qquad (X.6)$$

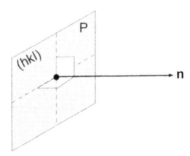

Table X.20 - Characterization of a plane P by its indices (hkl) or by its normal $\mathbf{n}$, the polarity of which is chosen arbitrarily.

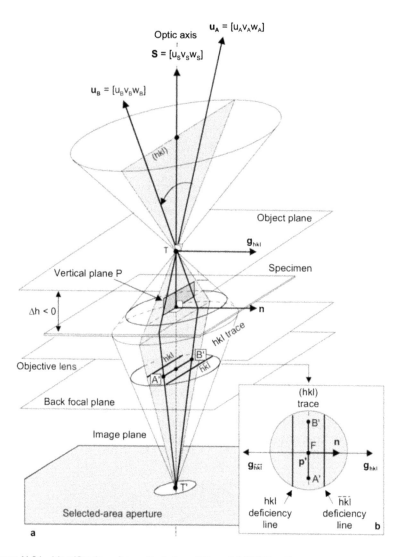

Figure X.21 - Identification of a vertical plane P from LACBED patterns.

a -The cross-product $u_A \times u_B$ of the two lattice directions u_A and u_B corresponding to the points A' and B' situated on the plane trace p' gives a reciprocal vector g_{hkl} perpendicular to u_A and u_B and to the set of lattice planes (hkl) containing the plane under study. As a result, $n = g_{hkl}$.

b - Relative arrangement of the pair of $h_1 k_1 l_1 / \overline{h_1} \overline{k_1} \overline{l_1}$ Bragg lines, the points A' and B' located on the plane trace p', the plane normal n and the reciprocal lattice vectors g_{hkl} and $g_{\overline{h}\overline{k}\overline{l}}$

Several methods can be used to identify the (hkl) Miller indices of a plane [X.2]. Two of them are described here. They require that the plane under study be parallel to the incident beam (the plane is vertical in the microscope). Its trace p' on the LACBED pattern is thus a line and all the artefacts that might arise from the observation of specimens with non-parallel faces are eliminated.

X.5.2.1 - First method P₁

This method is similar to the method L_2. Two points A' and B' situated on the trace p' of the plane are selected. They correspond to two crystallographic directions $\mathbf{u_A} = [u_A v_A w_A]$ and $\mathbf{u_B} = [u_B v_B w_B]$. According to figure X.21, the cross-product of these two directions $\mathbf{u_A} \times \mathbf{u_B}$ gives a reciprocal vector $\mathbf{g_{hkl}}$ perpendicular to both the $\mathbf{u_A}$ and the $\mathbf{u_B}$ directions and to the set of lattice planes (hkl) containing the plane P. The indices (hkl) of the plane under study are then obtained directly.

Note that the points A' and B' must have the configuration shown on figure X.21 to ensure that the cross product gives $\mathbf{g_{hkl}}$ and not $\mathbf{g_{\bar{h}\bar{k}\bar{l}}}$.

This method can be simplified by using zone axes for the $\mathbf{u_a}$ and $\mathbf{u_b}$ directions (this simplification is used in the following application).

The example of the stacking fault bordered by the two partial dislocations studied in the preceding paragraph is shown on figure X.22. The stacking fault is vertical for the two $[\bar{1}12]$ and $[\bar{2}13]$ zone axes (Figure X.22b). Their cross-product $[\bar{2}13] \times [\bar{1}12]$ gives $(\bar{1}1\bar{1})$ for the indices of the stacking-fault plane.

X.5.2.2 - Second method P₂

The second method consists in finding a pair of Bragg lines hkl / $\overline{hkl}$ parallel to the trace p' of the vertical plane. In agreement with figure X.21, the indices (hkl) of the plane are the same as the indices of the hkl Bragg line.

For example, on figure X.22, the stacking fault trace is parallel to the $1\bar{1}1$ and $\bar{1}1\bar{1}$ Bragg lines and its normal $\mathbf{n}$ is along the reciprocal vector $\mathbf{g_{1\bar{1}1}}$. The stacking-fault plane is thus $(1\bar{1}1)$.

More specific methods based on CBED and LACBED patterns have also been proposed for identifying the direction of a tilted stacking fault [X.3].

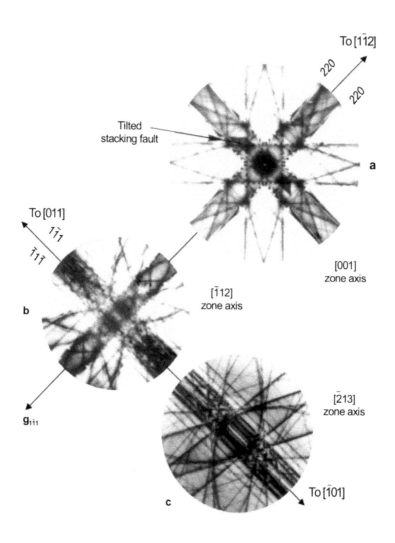

Figure X.22 - Identification of the indices (hkl) of the stacking fault bordered by the two partial dislocations n_a and n_a studied on figure X.14.

a - [001] zone-axis LACBED pattern. The stacking fault is tilted.

b, c - [$\bar{1}$12] and [$\bar{2}$13] LACBED patterns. The stacking fault is vertical and its trace is parallel to the $11\bar{1}$ / $\bar{1}11$ pair of Bragg lines.

Silicon specimen. Courtesy of P.H. Albarède.

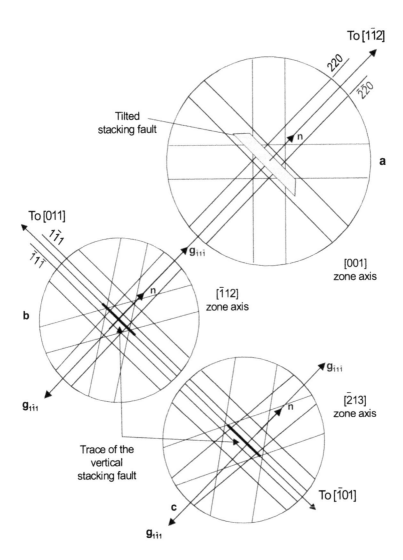

Figure X.22 - Continuation.
Interpretation of figures X.22b and c. The cross-product of the zone-axes $[\bar{1}12]$ and $[\bar{2}13]$ for which the stacking-fault plane is vertical allows the $(1\bar{1}\bar{1})$ stacking-fault plane to be identified.

The stacking fault normal is directed along the $g_{1\bar{1}\bar{1}}$ reciprocal vector implying that the fault plane is $(\bar{1}1\bar{1})$.

X.5.3 - Advantages and disadvantages of trace analyses from LACBED patterns

The absence of rotation between the image and the diffraction pattern is the main advantage of trace analyses from LACBED patterns.

The method has some disadvantages. It requires tilting the specimen for the image of the defect analysed to be contrasted. Difficulties are also encountered with bent specimens where the identification of Bragg lines parallel to the defect trace can be delicate. Indeed, a slight specimen deformation can produce a strong distortion (translation and rotation) of the Bragg lines, whereas the image of the defects is hardly not affected. This explains why the Bragg lines present on figures X.14 and X.15 are not always perfectly parallel to the defect trace. These considerations must be borne in mind for defect analysis from LACBED patterns and unbent areas of the specimens must be chosen.

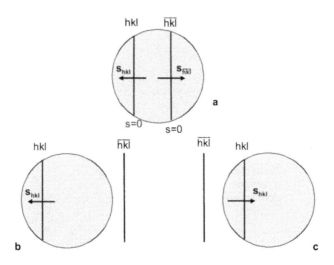

Figure X.23. Identification of the sign of the deviation parameter in a bright-field LACBED pattern.

a - The pattern contains the pair of hkl/$\overline{hkl}$ Bragg deficiency lines. The positive values of s point towards the outside of the pattern.

b, c - The LACBED pattern only contains one of the two lines of the pair. The positive value of s depends on the position of the second line.

X.6 - Identification of the deviation parameter s

In the chapter XI dealing with applications, we indicate that the determination of the Burgers vector **b** of dislocations requires a knowledge of the sign of the deviation parameter s. This parameter is zero on the Bragg line and positive or negative along parallel lines (see paragraph VI.5). When a bright-field LACBED pattern contains the pair of hkl/h̄k̄l̄ deficiency lines, the determination of the sign of the deviation parameter is straightforward: the positive values point towards the outside of the pattern (Figure X.23a). On the other hand, if the pattern contains only one of the two lines, the sign of s is less easy to establish. It depends on the position of the other line of the pair (Figures X.23b and c). This identification can be performed automatically with the "Electron Diffraction" software [VI.3]

Two examples are given on the simulations of figures X.8 and X.12 where the positive sign of s is indicated by means of a small line perpendicular to the Bragg line.

.

CHAPTER XI
Characterization of crystal defects from LACBED patterns

LACBED patterns are very well adapted to the analysis of crystal defects for the following reason: they are performed with a defocused incident beam and therefore contain information on both the reciprocal space (the Bragg lines) and the direct space (the shadow image of the illuminated area of the specimen). From that point of view, a LACBED pattern can be regarded as an image-diffraction mapping. In this chapter we show that crystal defects modify this image-diffraction mapping and produce typical effects, which can be used to identify them. Two other LACBED properties support this analysis. The first concerns the possibility of obtaining two-beam conditions (defects are generally most visible under these conditions) even with high hkl index reflections. In this case, disturbances from the other beams become negligible. The second property relates to the beneficial effect of angular filtering by the selected-area aperture, most of the inelastic electrons being removed, resulting in LACBED patterns of excellent quality.

In this chapter, we describe the characterization of crystal defects according to the defect dimension: point defects (0D), linear defects (1D) and planar defects (2D).

Examples will be taken mainly from defects found in face-centred cubic (fcc) structures. The results given here can hence be directly applied to many metals and alloys (gold, copper, aluminium, γ-iron...) and to many semiconductors (silicon, gallium, germanium...). They can be easily extended to other types of structures.

I.1 - Characterization of point defects

XI.1.1 - Crystallography of point defects

Point defects are crystal defects, the effect of which extends to only a few atoms around them. Some are shown on figure XI.1.

These are:

- vacancies. Individual atoms of the structure are missing. A relaxation occurs around these missing atoms involving small displacements of the neighbouring atoms.

- interstitial point defects. Atoms – usually small-sized foreign atoms – are inserted into the interstitial sites of the structure. Large local distortions may occur around these interstitials. Auto-interstitials are also encountered.

- substitutional point defects. Foreign atoms with a smaller or a larger radius than those of the matrix are exchanged with the matrix atoms.

- Shottky and Frenkel point defects. These are point defects observed in ionic crystals where electric neutrality must be preserved.

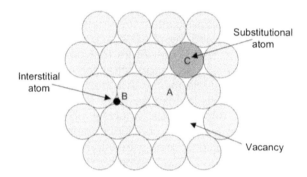

Figure XI.1 - Schematic description of some point defects.
- Vacancy. A matrix atom A is missing.
- Interstitial point defect. A foreign atom B is inserted into an interstitial site.
- Substitutional point defect. A foreign atom C replaces a matrix atom A.

XI.1.2 - Effect of point defects on LACBED Patterns

Point defects are too well localized in the crystal to be detected individually on LACBED patterns. Nevertheless, we can observe some induced effects connected with atomic relaxations when the density of point defects is high enough. We give here some qualitative results from examinations of three $Pd_{90}Pt_{10}$ alloys doped with tritium and aged for one, two and three months [XI.1]. After this thermal treatment, the three specimens contain increasing amounts of point defects, which eventually form aggregates and small helium bubbles produced by the decomposition of tritium.

The [3 4 15] zone-axis LACBED patterns shown on figure XI.2a were obtained from these three specimens with a cooling specimen-holder in order to reduce the effect or thermal vibrations. Careful examination shows that the intensity of the sharp and weak deficiency lines present in the central area of these patterns decreases and eventually cancels as the point disorder increases. These deficiency lines correspond to the largest Bragg angles visible on the pattern. We also note that the background intensity increases with ageing time. As reported in reference [XI.2], this point disorder may be considered as contributing a static term to the Debye-Waller factor.

The [3 4 15] dynamic LACBED simulations given on figure XI.2b reproduce very well the above experimental patterns. The parameter of the simulation is the Debye-Waller factor, which is given values between 0.001 and 0.005 nm². It is clearly seen that an increase of the Debye-Waller factor affects significantly the intensity of the reflections with large Bragg angle.

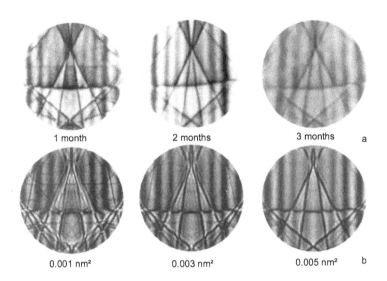

| 1 month | 2 months | 3 months | a |

| 0.001 nm² | 0.003 nm² | 0.005 nm² | b |

Figure XI.2 - Effect of point defects on the central area of LACBEB patterns.
a - [3 4 15] zone axis LACBED patterns from $Pd_{90}Pt_{10}$ alloys doped with tritium and aged for one, two and three months. The intensity of the sharp lines present in the central area of these patterns decreases with the ageing time. On the contrary, the background intensity increases with ageing.
b - Dynamic simulations of the [3 4 15] zone axis LACBED pattern obtained with three different Debye-Waller factors. The sharp Bragg lines and the background display modifications similar to those observed on the experimental patterns.
$Pd_{90}Pt_{10}$ specimen. Courtesy of CEA and B. Décamps.
Dynamic simulations obtained wth the Jems sofware from P. Stadelmann [X.1].

From the point of view of phenomenology, an increase of the point-defect density affects diffracted intensities in the same way as the temperature.

This type of analysis is only qualitative. A quantitative analysis is much more complex to carry out.

Note
This analysis could also be made from the excess lines (or deficiency lines) belonging to the high-order Laue zones on zone-axis CBED patterns, since these lines have the largest Bragg angles. Nevertheless, the CBED experiments are much more difficult to perform than the LACBED ones because they require perfect alignment of the incident beam with respect to the zone axis. In addition, specimen contamination by the focused incident beam and the absence of inelastic filtering make it impossible to observe very weak Bragg lines.

XI.2 - Characterization of linear defects: dislocations

XI.2.1 - Crystallography of dislocations

A dislocation is a linear defect (1D defect), involved in the plastic deformation of crystals. It consists of a strained crystal zone located around a **dislocation line.** This zone can move through the crystal as a consequence of applied stress (the dislocation slips) or thermal effects (the dislocation climbs). When the dislocation line leaves the crystal, a permanent deformation characterized by a vector **b** called the **Burgers vector** is produced. This Burgers vector is a lattice direction:

$$\mathbf{b} = [u_b v_b w_b] = u_b \mathbf{a} + v_b \mathbf{b} + w_b \mathbf{c} \qquad (XI.1)$$

Three types of strained zones can be distinguished:

- strained zones resulting from the insertion, like an "edge", of an additional atomic plane above or below the dislocation line (noted T or ⊥). For this reason, the corresponding dislocations are called **edge dislocations** (Figure XI.3a and b).

- strained zones, such as those shown on figure XI.3c and d, where the atomic planes perpendicular to the dislocation line are transformed into a helical surface. These dislocations are called **screw dislocations**.

- strained zones made up of a mixture of the two previous types. They are called **mixed dislocations**.

XI.2.1.1 - Burgers circuit

A Burgers circuit is used to characterize the Burgers vector **b** of a dislocation. In order to avoid any ambiguity about the sign of **b**, the FS/RH (Finish to Start; Right-Handed) convention is applied.

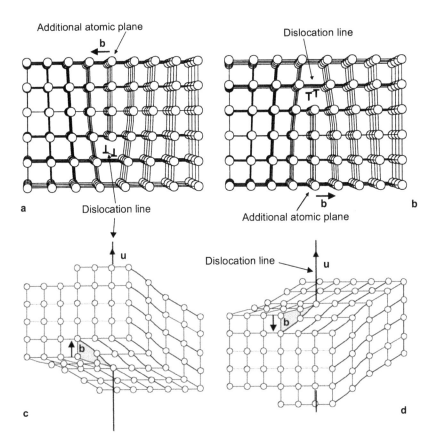

Figure XI.3 - Schematic description of edge and screw dislocations.
a, b - Positive and negative edge dislocations. The two dislocations are formed by the insertion like an "edge" of an atomic plane above or below the dislocation line.
The vector **u** characterizing the dislocation line points into the page.
c, d - Right and left screw dislocations. The planes perpendicular to the dislocation line are transformed into helical surfaces with a right- or left-handed helix.

A closed loop is drawn in the right-handed orientation (RH), around the dislocation line (Figure XI.4a). The same circuit in a perfect crystal (Figure XI.4b) displays a closure gap typical of the Burgers vector. Its orientation is defined by the FS convention (from the final point F towards the starting point S).

Note that the orientation of **b** depends on the polarity of the dislocation line. This polarity is arbitrarily defined in terms of a lattice vector $\mathbf{u} = [u_u v_u v_u] = u_u \mathbf{a} + v_u \mathbf{b} + w_u \mathbf{c}$.

Edge dislocations are said to be positive or negative when the additional atomic plane is located above or below the dislocation line respectively (Figures XI.3a and b). Right- or left-hand screw dislocations are defined according to the right or left character of the helical sequence of planes perpendicular to the line (Figures XI.3c and d).

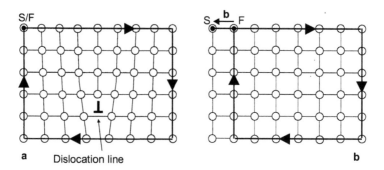

a   Dislocation line   b

Figure XI.4 - Burgers circuit around an edge dislocation.
a - A closed right-handed circuit is drawn around the dislocation line.
b - The same circuit is drawn in a perfect crystal. The F → S gap characterizes the Burgers vector **b**.
On this diagram, the vector **u** points into the page.

From the point of view of the Burgers vector:
    - an edge dislocation is characterized by a vector **b** perpendicular to the dislocation line **u** (Figure XI.5a),
    - a screw dislocation is characterized by a vector **b** parallel to the dislocation line **u** (Figure XI.5b),
    - a mixed dislocation is characterized by a vector **b** at an arbitrary angle to its line. In this case, the Burgers vector **b** can be decomposed into edge and screw components, b_c and b_v (Figure XI.5c).

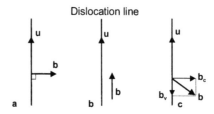

Figure XI.5 - Description of the three types of dislocations.
a - Edge dislocation. The Burgers vector **b** is perpendicular to the dislocation line **u**.
b - Screw dislocation. The Burgers vector **b** is parallel to the dislocation line **u**.
c - Mixed dislocation. The Burgers vector **b** has an unspecified orientation with respect to **u**. It can be decomposed into edge and screw components, b_c and b_v.

For energetic reasons, Burgers vectors are usually the shortest lattice translation vectors (for example, **b** = 1/2<110> in fcc structures). The dislocations are then called **perfect dislocations**.

In some structures, (for example, the face-centred cubic, the hexagonal close-packed and the body-centred cubic structures), a perfect dislocation **b** can dissociate into two **partial dislocations** with Burgers vectors $\mathbf{b_1}$ and $\mathbf{b_2}$ smaller than the lattice translation vectors and so that $\mathbf{b_1} + \mathbf{b_2} = \mathbf{b}$. Since these two partial dislocations are always separated by a stacking fault, they will be described in the paragraph XI.3.2 dealing with stacking faults.

In conclusion, the full characterization of a dislocation requires a knowledge of its two vectors **b** and **u**.

XI.2.2 - Identification of the Burgers vector from LACBED patterns

In the strained crystal zone around a dislocation line, the atoms are displaced with respect to their normal position in a perfect crystal. Obviously, this strained zone produces a local modification of the lattice plane orientations and therefore of the corresponding diffraction conditions. We indicated previously that LACBED patterns could be considered as diffraction-image mappings. We can expect that these mappings will be disturbed by the presence of dislocations.

This was confirmed when Cherns and Preston [i.7] reported, in 1986, that the strain field around dislocations distorts the Bragg lines, producing typical splittings (Figure XI.6). A Bragg line $\mathbf{g_{hkl}}$ simultaneously rotates and separates into a system of fringes at the intersection with the trace **u'** of a dislocation line **u**. The interfringe number n is directly related to the Burgers vector by the dot product:

$$\mathbf{g_{hkl} \cdot b} = (h\mathbf{a^*} + k\mathbf{b^*} + l\mathbf{c^*}) \cdot (u_b\mathbf{a} + v_b\mathbf{b} + w_b\mathbf{c}) = hu_b + kv_b + lw_b = n \qquad (XI.2)$$

Two different splitting arrangements can be observed for a given n value (Figure XI.7a and b). They are mirror related and depend on the relative orientation of the deviation parameter s with respect to the dislocation line **u**. The sign of n can therefore be identified by using the Cherns and Preston rules given in figures XI.7a and b [XI.3].

The interpretation of these experimental results was confirmed by simulating theoretical LACBED patterns for various types of dislocations by using the kinematical and dynamical theories. Examples corresponding to the effects of a screw dislocation on Bragg lines with n = 0, ±1, ±2 and ±3 are given on figure XI.8. Many other examples can be found in references [XI.4 and 5].

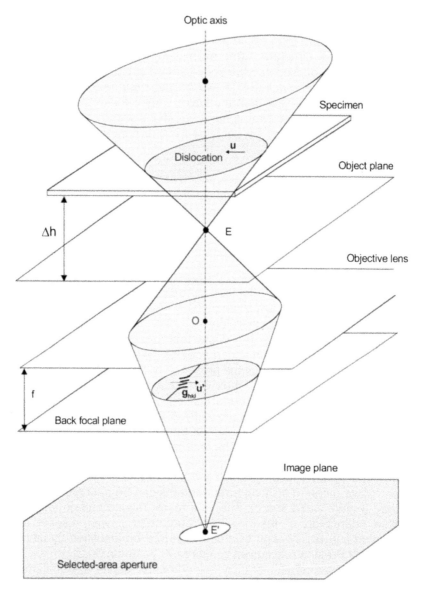

Figure XI.6 - Effect of a dislocation on LACBED patterns.
The Bragg line g_{hkl} undergoes a rotation and separates into a fringe system at the intersection with the trace u' of a dislocation line u. The interfringe number n is directly related to the dot product $g_{hkl}.b = n$. On this diagram n = 3.

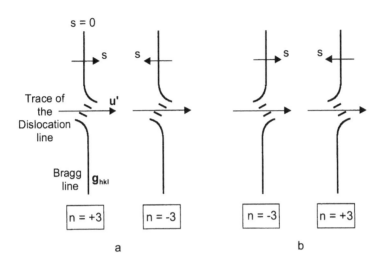

Figure XI.7 - Cherns and Preston rules [XI.3]. The sign and value of n can be identified from these rules. Note that the rules require a knowledge of the positive direction of the deviation parameter s.

XI.2.2.1 - Experimental identification of the Burgers vector b

The Cherns and Preston rules are the basis of the experimental identification of the Burgers vector **b** from LACBED patterns.

The method consists in observing at least three splittings in order to establish the three following equations:

$$\mathbf{g}_{h_1k_1l_1}.\mathbf{b} = h_1u_b + k_1v_b + l_1w_b = n_1 \qquad (XI.3)$$
$$\mathbf{g}_{h_2k_2l_2}.\mathbf{b} = h_2u_b + k_2v_b + l_2w_b = n_2$$
$$\mathbf{g}_{h_3k_3l_3}.\mathbf{b} = h_3u_b + k_3v_b + l_3w_b = n_3$$

The solution of this linear system gives the indices u_b, v_b and w_b of **b**. The method requires the identification of the hkl indices of the three Bragg lines involved in the analysis as well as the positive direction of their corresponding deviation parameter s. We have indicated how to identify these parameters in chapter X.

Note
In order to be able to solve the above simultaneous equations, the three $\mathbf{g}_{h_1k_1l_1}$, $\mathbf{g}_{h_2k_2l_2}$ and $\mathbf{g}_{h_3k_3l_3}$ Bragg lines must not originate from the same zone axis.

Depending on the length of the trace of the dislocation line on the pattern, two experimental procedures can be used to establish the three equations required for the analysis.

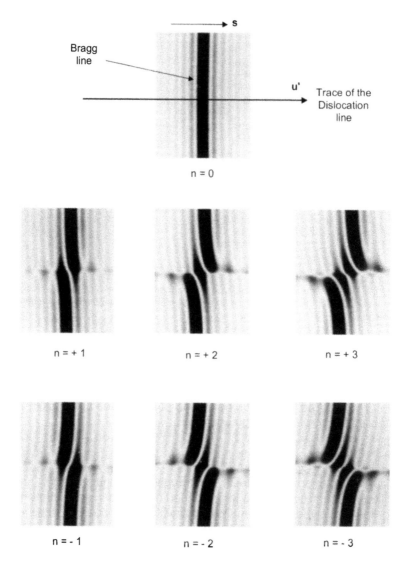

Figure XI.8 - Kinematical simulations of the splittings observed at the intersection of a screw dislocation line with some Bragg lines whose dot product $g_{hkl}.b$ is equal to 0, $\pm$ 1, $\pm$ 2 and $\pm$ 3.

The screw dislocation is located at the centre of a specimen with thickness t = 100 nm. Courtesy of A.R. Preston and D. Cherns.

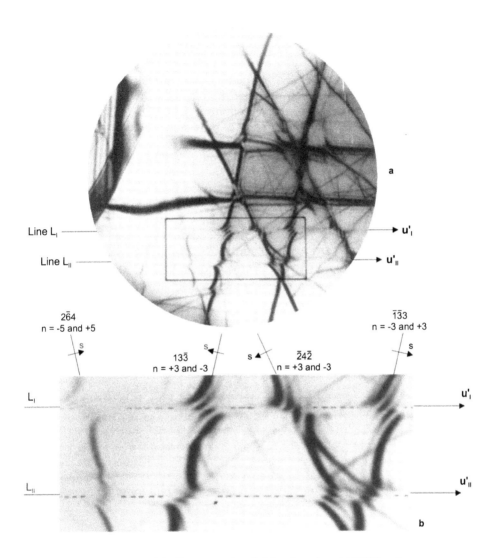

Line L_I ⟶ u'_I

Line L_II ⟶ u'_II

a

$2\bar{6}4$
n = -5 and +5

$\bar{1}33$
n = -3 and +3

$13\bar{3}$
n = +3 and -3

$\bar{2}4\bar{2}$
n = +3 and -3

L_I   u'_I

L_II   u'_II

b

Figure XI.9 - Identification of the Burgers vector of dislocations from LACBED patterns.
a - The LACBED pattern contains the traces of two long dislocation lines L_I and L_II intersecting several Bragg lines.
b - Enlargement of the previous pattern. The indexing of this pattern is given on figure X.8. The Burgers vectors $\mathbf{b}_I = \frac{1}{2}[0\bar{1}1]$ and $\mathbf{b}_{II} = \frac{1}{2}[0\bar{1}1]$ are identified for the lines L_I and L_II respectively. Silicon specimen deformed at high temperature. Courtesy of J.P. Michel.

- If the trace of the dislocation line is long enough, it can intersect three Bragg lines (or more) simultaneously. This is what happens for the two dislocation lines L_I and L_{II} present on figure XI.9. This situation is ideal since a single LACBED pattern allows the Burgers vector to be identified.
- If the trace of the dislocation line is short, the line must be moved by shifting or tilting the specimen in order to make it intersect at least three Bragg lines. Three different LACBED patterns are then required. The three patterns on figure XI.10 illustrate this case.

XI.2.2.1.1 - First example of Burgers vector identification

This first example is taken from a silicon specimen deformed at high temperature (Figure XI.9). The indexing of the corresponding LACBED pattern has already been given in paragraph X.4.1 and on figure X.8. The traces of two very long dislocation lines L_I and L_{II} are present on this pattern and intersect many Bragg lines. The intersections with the Bragg lines $\overline{2}6\overline{4}$, $13\overline{3}/\overline{1}3\overline{3}$ and $\overline{2}4\overline{2}$ produce well-defined splittings whose interfringe numbers n are given in Table XI.1.

g_{hkl}	n_{exp} for Line L_I	n_{exp} for Line L_{II}
$\overline{2}6\overline{4}$	-5	+5
$13\overline{3}$	+3	-3
$\overline{2}4\overline{2}$	+3	-3
	$b_I = \frac{1}{2}[01\overline{1}]$	$b_{II} = \frac{1}{2}[0\overline{1}1]$

Table XI.1 - Experimental n values for the splittings observed at the intersection of the dislocation lines L_I and L_{II} with three Bragg lines of the LACBED pattern on figure XI.7. The Burgers vectors $b_I = \frac{1}{2}[01\overline{1}]$ and $b_{II} = \frac{1}{2}[0\overline{1}1]$ are identified from this table for the two dislocation lines L_I and L_{II}.

Three linear equations can be established for each of the L_I and L_{II} dislocation lines. For the first line L_I, we obtain:
2u - 6v + 4w = -5
u + 3v - 3w = 3
- 2u + 4v - 2w = 3

They lead to the Burgers vector $b_I = \frac{1}{2}[01\overline{1}]$. For the second dislocation line L_{II}, we obtain in the same way, $b_{II} = \frac{1}{2}[0\overline{1}1]$. Both of them are perfect dislocations with opposite Burgers vectors.

XI.2.2.1.2 - Second example of Burgers vector identification

This second example concerns a garnet specimen deformed at high temperature and containing short dislocation lines. The previous identification method can no longer be used. Here, the dislocation studied must be moved and placed on at least three Bragg lines, as shown on the patterns of figure XI.10. The indexing of these patterns has already been given in paragraph X.4.2 and on figure X.12. The n values, listed in table XI.2, allow us to identify a perfect dislocation with a Burgers vector $b = \frac{1}{2}[\bar{1}\bar{1}1]$ [XI.6].

g_{hkl}	n_{exp}
$010\bar{2}$	-4
$\bar{1}\bar{4}3$	+4
$\bar{1}94$	-2
	$b = \frac{1}{2}[\bar{1}\bar{1}1]$

Table XI.2 - Experimental n values for the splittings observed at the intersection of a dislocation line with three Bragg lines present on the LACBED patterns on figure XI.10. The Burgers vector $b = \frac{1}{2}[\bar{1}\bar{1}1]$ is identified from this table.

XI.2.2.1.3 - Third example of Burgers vector identification

This last example concerns a Cu-Zn-Al shape-memory alloy characterized by a very high anisotropy factor $2C_{44} / (C_{11} - C_{12}) = 12$ [XI.7].

The determination of the Burgers vector of a dislocation in conventional transmission electron microscopy is generally a very delicate task for anisotropic materials. The two invisibility criteria $g_{hkl}.b = 0$ and $g_{hkl}.(b \times u) = 0$ cannot be applied with confidence owing to the systematic presence of some residual contrast.

Anisotropy also has an effect on the dislocation splittings observed on LACBED patterns [XI.8]. Some of them are strongly distorted (Figure XI.11d) while other display a nearly normal aspect (Figures, XI.11a, b and c). For these unaffected splittings, the interfringe number n can be identified reliably and the Cherns and Preston rules applied. Therefore, it is possible to analyse dislocations in these highly anisotropic materials provided the unaffected splittings are considered and the distorted ones rejected for the analysis. This is not a difficult task since many splittings are usually available.

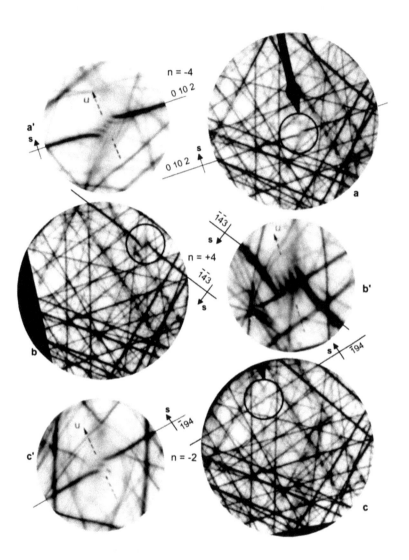

Figure XI.10 - Identification of the Burgers vector of a dislocation present in a garnet specimen deformed at high temperature.
a, b, c - A short trace of a dislocation line **u** intersects the three Bragg lines 0 10 2 (a and a'), $\overline{1}4\overline{3}$ (b and b') and $\overline{1}94$ (c and c'). The indexing of these patterns is given on figure X.12. The Burgers vector **b** = ½ [$\overline{11}$1] is identified from these three LACBED patterns.
Garnet LACBED patterns. Courtesy of P. Cordier.

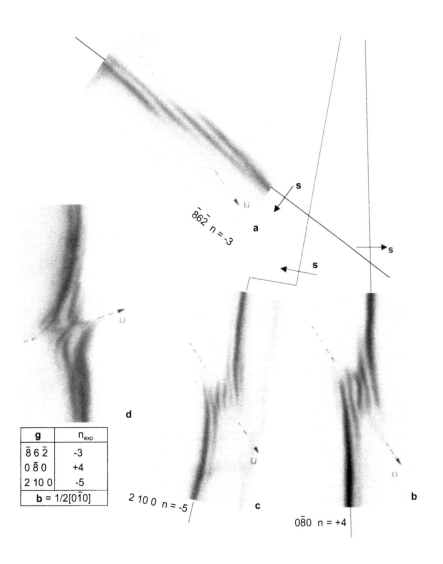

g	$n_{exp.}$
$\bar{8}\,6\,\bar{2}$	-3
$0\,\bar{8}\,0$	+4
$2\,10\,0$	-5
$\mathbf{b} = 1/2[0\bar{1}0]$	

Figure XI.11 - Identification of the Burgers vector of a dislocation in anisotropic materials.
a, b, c - Aspects of three splittings almost unaffected by anisotropy. They allow the Burgers vector to be identified as $\mathbf{b} = \frac{1}{2}[0\bar{1}0]$.
d - Aspect of a splitting strongly distorted by anisotropy.
Cu-Zn-Al shape-memory alloy. Courtesy of J. Pons and R. Portier.

For example, the three unaffected splittings $\overline{8}6\overline{2}$, $0\overline{8}0$ and 2 10 0 on figure XI.11 can be used to identify the Burgers vector **b** = ½[$0\overline{1}0$] in this shape-memory alloy.

The possibility of characterizing the Burgers vectors in anisotropic materials from LACBED patterns is a major advantage of this method over conventional methods.

XI.2.2.2 - Operating conditions for the Burgers vector identification

The identification of the Burgers vector from LACBED patterns requires clear splittings from which the interfringe number n can be counted reliably.

The best splittings are obtained when:
- the middle of the trace of the dislocation line (the part of the dislocation line located in the middle of the thin foil) intersects the Bragg line. This means that the dislocation line must be accurately positioned with respect to the Bragg line.
- the dislocation line is tilted with respect to the electron beam. Vertical lines are not suitable. Horizontal lines are also not suitable unless they are located near the middle of the thin foil.

XI.2.2.2.1 - Selection of the Δh value

The main experimental variable is the value of Δh, i.e. the specimen height with respect to the object plane of the objective lens. In chapter VI, we showed that a variation of the specimen height affects the shadow image but not the diffraction pattern itself. This means that the deviation parameter s remains unchanged but the magnification of the strained zone surrounding the dislocation line decreases as Δh increases. The splitting spreads out as shown on figures XI.12a, b and c when Δh decreases. It is advisable to select the Δh value that gives the best splitting visibility.

Since the image suffers a 180° rotation between positive and negative values of Δh (see paragraph VI.6.5), the splitting transforms into its mirror image when the sign of Δh changes (Figures XI.12d and e). **It is thus very important to ensure that all the LACBED patterns used for a Burgers vector identification are made with the same sign of Δh.**

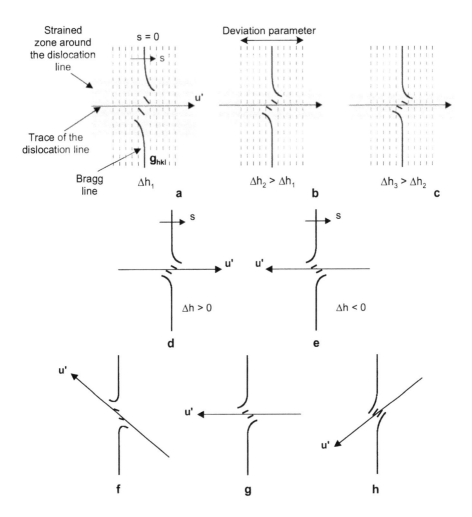

Figure XI.12 - Appearance of the splittings observed at the intersection of the trace **u'** of a dislocation line **u** with a Bragg line.

a, b, c - Effect of the value of Δh. The magnification of the strained zone around the dislocation line is modified but not the deviation parameter. The fringes spread out as Δh decreases.

d, e - Effect of the sign of Δh. The image suffers a 180° rotation between the positive and negative Δh values. The splitting is transformed into its mirror image.

f, g, h - Effect of the orientation of the trace **u'** of the dislocation line with respect to the Bragg line. The best splitting is obtained when the trace is perpendicular to the Bragg line.

XI.2.2.2.2 - Choice of the Bragg lines

The choice of the Bragg lines is essential. The best splittings are usually observed with quasi-kinematical Bragg lines, i.e. lines displaying a single minimum or maximum. The selection of n is also important for reliable determination of the value of n; values of n higher than 6 or 7 can be difficult to identify.

Dynamical lines made up of a set of fringes are not very suitable because they give splittings whose appearance is very complex. Their interfringe number n is very difficult, even impossible, to identify.

The identification methods described above are performed on non-zone-axis LACBED patterns. A zone-axis pattern is not very useful since the Bragg lines present on it are generally too dynamical to determine n. Moreover, the Bragg lines implied in the analysis of a dislocation should not all come from the same zone axis, because it would not be possible to determine **b**.

It is also advisable to choose Bragg lines making an angle as close as possible to 90° with the trace of the dislocation line in order to obtain good splittings. The splittings become distorted when this angle is far from 90° (Figures XI.12f, g and h).

XI.2.2.2.3 - Choice of the probe size

A small probe size improves the image resolution but decreases the brightness of the pattern. With very small probe sizes, it is preferable to work with a CCD camera in order to improve the fringe visibility.

XI.2.2.2.4 - Choice of the technique

A dislocation is only visible under strong diffraction conditions, which implies that it is only well contrasted when it intersects Bragg lines. Everywhere else, it is not contrasted and therefore invisible. This create serious problems when a dislocation line must be moved from one Bragg line to another. During this displacement, the image of the dislocation disappears and the dislocation can be lost. For an isolated dislocation or for a dislocation located in an easily recognizable specimen area, this lack of contrast is not a real problem and the LACBED technique is then recommended because it rapidly produces the best patterns.

On the contrary, if the dislocation is close to many other dislocations or if there is no nearby specimen feature for locating it reliably, the LACBED technique can prove to be very laborious. An alternative solution is to use techniques that maintain the specimen eucentricity, i.e. the defocus CBED, the eucentric LACBED and the CBIM techniques. First, the dislocation under study is accurately placed at the eucentric height in the centre of the screen. Then, the specimen can be tilted in order to move the dislocation line from one intersection with a Bragg line to another without any risk of losing it.

XI.2.2.2.5 - Choice of the bright- or dark-field patterns

Bright-field patterns have the advantage of displaying simultaneously many Bragg lines, which facilitates the indexing of the Bragg lines required by the analysis. Dark-field patterns are more contrasty than bright-field patterns so that the n value can be obtained more reliably (figure XI.13d). Nevertheless, the problem mentioned in the previous paragraph concerning the absence of dislocation contrast is more crucial for these dark-field patterns and it can be very difficult to position a dislocation line on an excess line. In addition, it is also necessary to take into account the image shift between bright- and dark-field patterns. This problem can be solved in the following way. Imagine that we want to observe a splitting at the intersection of a dislocation line with an hkl excess line on a dark-field LACBED pattern.

- First, the bright-field pattern is observed and the dislocation line is placed on the hkl deficiency line so that a splitting is observed (Figure XI.13a).

- In a second step, the $\overline{hkl}$ deficiency line is placed at the centre of the screen using the post-specimen beam deflection coils (Figure XI.13b).

- Lastly, the dark-field pattern is observed (Figure XI.13b). The hkl excess line and the splitting are located at the centre of the screen. Note that the same splitting – and not the mirror related spiltting – is observed for the hkl excess and deficiency Bragg lines.

XI.2.2.2.6 - A few difficulties

The main difficulty encountered during the analysis of Burgers vectors from LACBED patterns arises with specimens with a high dislocation density. In this case, the dislocation strain fields may overlap and good splittings are difficult to observe. It is estimated that two dislocations can be characterized when they are separated by a distance greater than fifty nanometres. Despite this limitation, it remains possible to observe good splittings even when the dislocation density is high. We recall that good splittings are only observed when the Bragg line intersects the middle of a dislocation line. With many dislocations, it might happen that this condition is satisfied for only one dislocation at a time.

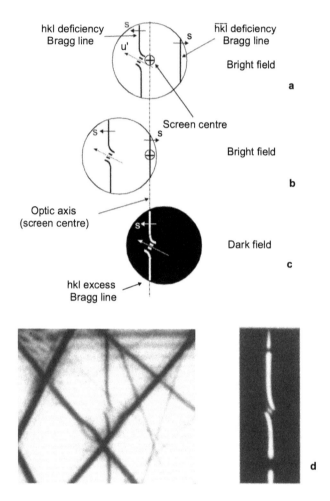

Figure XI.13 - Procedure used to observe splittings on dark-field LACBED patterns.

a - The dislocation line is placed on the hkl deficiency line of the bright-field pattern.

b - The whole pattern is shifted with the post-specimen beam deflection coils until the $\overline{hkl}$ deficiency line is located at the centre of the screen.

c - The hkl dark-field pattern is observed. The hkl excess line is at the centre of the screen and the splitting is observed.

d - Splittings observed at the intersection of a "stair-rod" dislocation with the 10 0$\overline{2}$ deficiency and excess lines on bright- and dark-field LACBED patterns. Note that the same splitting is observed for the deficiency and excess lines but the latter has a better contrast.

Silicon specimen. Courtesy of P. H. Albarède.

XI.2.2.3 - Applications of the LACBED technique to the characterization of various types of dislocations

XI.2.2.3.1 - Applications to electron-beam-sensitive crystals

Electron-beam-sensitive materials represent a very interesting application. The LACBED method is a defocused method, which means that the convergent incident beam illuminates a relatively large specimen area. Consequently, each point of the specimen receives a much lower electron dose than in the CBED or Kossel techniques, where the whole incident beam is focused on a small specimen area. The LACBED technique can therefore be applied to electron-sensitive crystals such as organic or mineral crystals [XI.9].

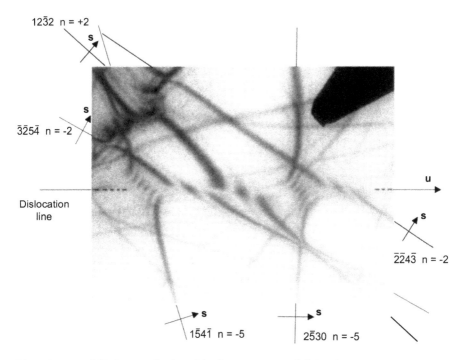

Figure XI.14 - LACBED identification of the Burgers vector of dislocations in electron-sensitive materials. Example of a quartz specimen observed close to the $[10\bar{1}2]$ zone axis. This LACBED pattern allows the Burgers vector to be identified as **b** = 1/3 $[\bar{1}2\bar{1}0]$ for a long dislocation line intersecting the whole pattern.
LACBED pattern from a quartz specimen. Courtesy of P. Cordier.

The example of quartz is shown on figure XI.14 [XI.10]. This mineral is very sensitive to the electron beam and is rapidly rendered amorphous so that the analysis of dislocations by conventional methods is very difficult or even impossible in this material. This obstacle no longer exists with the LACBED technique, with which a specimen can be observed for several hours without any beam damage.

XI.2.2.3.2 - Other applications

The LACBED method has been successfully applied to dislocations present in many materials: metals and alloys, semiconductors, ceramics, minerals, quasicrystals (figure XI.15). Details on these applications are given in the reference list at the end of the book.

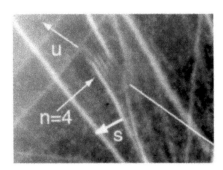

Figure XI.15 - LACBED identification of the Burgers vector of dislocations in quasicrystals. For these quasicrystals, the identification of the Burgers vector is more complex since at least six splittings must be observed.
Courtesy of D. Caillard and D. Gratias.

XI.2.2.4 - Advantages of the LACBED technique

The main advantages of this method over conventional methods are:

- the possibility of obtaining perfect two-beam conditions even with high hkl indices. To a large extent, the good quality of the splittings is related with this property.

- the high reliability of the Burgers vector identification, especially if many splittings are considered.

- the possibility of identifying very small Burgers vectors. Small Burgers vectors of about 0.05 nm can be analysed with this method [XI.11].

- the possibility of identifying Burgers vectors in anisotropic materials,
- the possibility of identifying Burgers vectors in beam-sensitive materials.

Unlike conventional method, where the identification is obtained from the absence of an effect (the specimen is tilted so that the dislocation becomes out of contrast), the LACBED method is based on the presence of an effect (the specimen is shifted or tilted so that a splitting is observed). From this point of view, the LACBED method is closer to the method of identification in which experimental and simulated images obtained in the two-beam conditions for different g_{hkl} vectors are compared [XI.12]. The main problem of this approach, again, lies in the difficulty of obtaining exact two-beam conditions.

Finally, it is unnecessary to make any assumption concerning the nature of the Burgers dislocation vector when using the LACBED method.

XI.2.3 - Identification of the dislocation line u

The full characterization of a dislocation requires, in addition to a knowledge of its Burgers vector **b**, a knowledge of the dislocation line, i.e. its vector **u**.

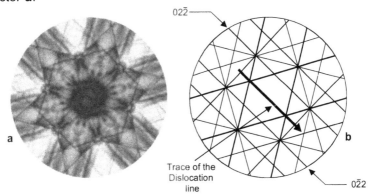

Figure XI.16 - Trace analysis of a dislocation line **u** from a LACBED pattern.
a - The trace **u'** of the dislocation line **u** crosses the central area of a zone-axis LACBED pattern. Note that the dislocation shears the LACBED pattern. This effect is typical of a screw dislocation (see chapter XI.2.4).
b - Interpretation of the experimental pattern. The trace **u'** is parallel to the $02\bar{2}$ and $0\bar{2}2$
Bragg lines, meaning that the dislocation line **u** is contained in the $(0\bar{2}2)$ lattice planes.
Ni$_3$Al specimen. Courtesy of E. Jezierska.

LACBED patterns are again well adapted to this identification since the shadow image and the diffraction pattern are present simultaneously. We can thus identify the vector **u** by trace analysis as described in chapter X.

On the example given on figure XI.16, the trace **u'** of the dislocation line **u** is parallel to the $0\bar{2}2$ and $02\bar{2}$ Bragg lines, meaning that the dislocation line is contained in the ($0\bar{2}2$) lattice planes. A second experiment with another crystal orientation is required to identify **u**.

XI.2.4 - Identification of the character of the dislocation

The identification of the edge, screw or mixed character of a dislocation requires a knowledge of both the Burgers vector **b** and the dislocation line vector **u** using the above methods.

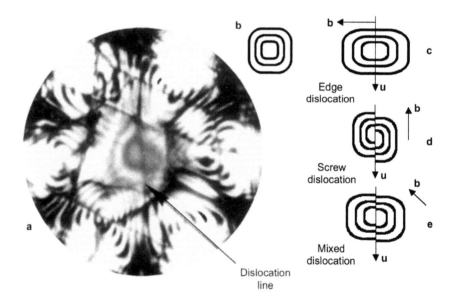

Figure XI.17 - Effect of a dislocation on the central area of a zone-axis LACBED pattern.
a - Mixed dislocation observed along the <111> zone axis.
b - Central area of the pattern.
c, d, e - Schematic descriptions of the effect of edge (c), screw (d) and mixed (e) dislocations on the central area of zone-axis LACBED patterns. Descriptions taken from Wen *et al.* [XI.13].
Silicon specimen. Courtesy of J. P. Michel.

Another very elegant way of detecting the dislocation character was proposed by Wen *et al.* [XI.13] and de Blasi *et al.* [XI.14]. It consists in observing the effect of a dislocation line on the central area of zone-axis LACBED patterns (figure XI.16a). The various configurations are typical of the edge, screw or mixed character of a dislocation and can be used to identify it (figure XI.17b).

In view of the close relationship between partial dislocations and stacking faults, methods of identifying the Burgers vector of partial dislocations will be described in paragraph XI.3.2, where stacking faults are examined. The identification of dislocations present at grain boundaries will be described in paragraph XI.3.6, where grains boundaries are considered.

XI.3 - Characterization of planar (2D) defects

LACBED is also very well adapted to the analysis of planar defects for the same reasons as those given for dislocations.
The main planar defects are stacking faults, antiphase boundaries and grain boundaries. We first consider stacking faults.

XI.3.1 - Characterization of stacking faults

XI.3.1.1 - Crystallography of stacking faults

Crystal structures are often described in terms of stacking sequences of atomic planes. This type of description is particularly useful for the face-centred cubic structure and the hexagonal close-packed structure of metals and alloys. Such structures are very conveniently represented in terms of stackings of compact atomic planes as shown on figure XI.18a. These compact planes are regularly stacked in specific sequences.
How can we describe these stacking sequences?
We consider a first compact plane (Figure XI.18b). The atoms in this plane are located at positions A (for the sake of clarity, the atoms are represented with a reduced size on this drawing).
The atoms of the second compact plane can occupy either of the two positions:
- positions B (Figure XI.18c). In this case, the atoms of this second layer are shifted, relative to the atoms of the first by one of the three equivalent vectors $+R_1$ or $+R_2$ or $+R_3$. The shift from a position A to a position B is thus characterized by a vector $+R$ ($+R_1$, $+R_2$, $+R_3$).

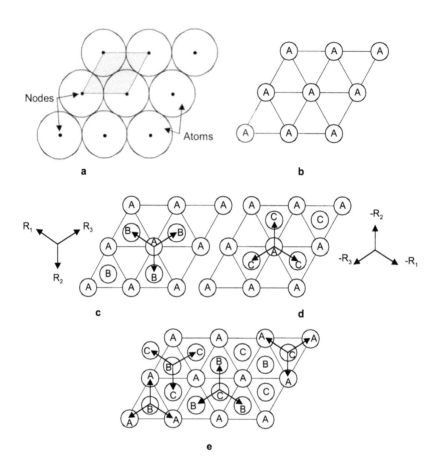

Figure XI.18 - Description of compact structures in terms of stacking sequences of compact atomic planes.

a - Description of a compact atomic plane. The atoms are contiguous and form a two-dimensional hexagonal lattice.

b - Schematic description of a compact atomic plane. The atoms are represented by small circles and occupy positions A.

c - Stacking of two successive compact planes. Here, the atoms of the second layer occupy positions B. The shift from a position A to a position B is described by one of the three equivalent vectors $+R_1$, $+R_2$ or $+R_3$.

d - The atoms of the second layer can also occupy positions C. The shift from a position A to a position C is described by the three equivalent vectors $-R_1$, $-R_2$ or $-R_3$.

e - The shift from a position B to a position C and from a position C to a position A is described by the vectors $+R_1$ or $+R_2$ or $+R_3$. The opposite shift from B to A and from C to B is described by the opposite vectors $-R_1$ or $-R_2$ or $-R_3$.

- or positions C (Figure XI.18d). In this case, the atoms of the second layer are shifted relative to the atoms of the first layer by a vector $-R_1$ or $-R_2$ or $-R_3$. The shift from a position A to a position C is thus described by a vector $-R$ ($-R_1$, $-R_2$, $-R_3$).

In the same way, a layer C is shifted by $+R$ and a layer A by $-R$ with respect to a layer B (Figure XI.18e). A layer A is shifted by $+R$ and a layer B by $-R$ with respect to a layer C (Figure XI.18e).

In conclusion, a shift from A → B → C → A...is characterized by $+R$, and a shift in the opposite direction, from A → C → B → A... by $-R$.

Note
The R vectors only represent the shift parallel to the compact planes. They do not take into account the spacing between the planes.

Any stacking sequences of these compact planes where the successive planes are shifted by $+R$ or $-R$ correspond to a compact structure made of identical atoms. An infinite number of compact structures can be generated by regular stackings of these planes. The two structures actually encountered correspond to the two simplest sequences:

- the binary sequence ABABAB... (or ACACAC...).describes **the hexagonal close-packed structure (hcp)** of metals. In this case, the compact planes are {0001} planes and the shifts R are of the type $1/3 < 10\bar{1}0 >$ (Figure XI.19a).

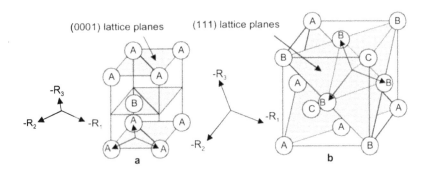

Figure XI.19 - Description of the hexagonal close-packed (hcp) and face-centred cubic (fcc) structures of metals.
a - Hexagonal close-packed structure. The stacking sequence is ABABAB... The compact planes are {0001} planes and the R vectors are of the type $1/3 < 10\bar{1}0 >$.
b - Face-centred cubic structure. The stacking sequence is ABCABC... The compact planes are {111} planes and the vectors R are of the type $1/6 < 112 >$.

- the ternary sequence ABCABC... (or ACBACB...). This sequence corresponds to **the face-centred cubic structure (fcc)**. The compact planes are {111} planes and the shifts **R** are of the type 1/6<112> (Figure XI.19b).

XI.3.1.1.1 - Description of the stacking sequences

In order to describe the stacking sequences, it is useful to represent the structures along a vertical cross-section containing one of the **R** shifts (for example, the shift R_1 along the line L'L as shown on figure XI.20a). The sequences ABABAB... and ABCABC... are clearly highlighted in this way on figures XI.20b and c.

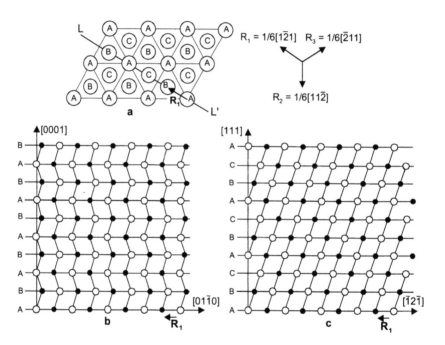

Figure XI.20 - Description of the stacking sequences of the hexagonal close-packed (hcp) and face-centred cubic (fcc) structures.
a - Projection of the A, B and C atomic positions.
b - Hexagonal close-packed structure. Vertical cross-section parallel to LL' including the vector R_1. The stacking sequence ABABAB... is clearly highlighted.
c - Face-centred cubic structure. Vertical cross-section parallel to LL' including the vector R_1. The stacking sequence ABCABCABC... is highlighted.
The black atoms are at z = 0, 1, 2.....and the white ones at z = 1/2, 3/2, 5/2...

Any modification of the normal stacking sequence constitutes a stacking fault. Two types of stacking faults are encountered: **intrinsic** and **extrinsic stacking faults**.

- Intrinsic stacking faults

We consider the fcc structure with the ABCABC...stacking sequence (Figure XI.22a). An intrinsic stacking fault can be created by two different mechanisms:

- by displacing, under the action of a shear stress, the lower part of the crystal by a shift $-R_1 = 1/6[12\bar{1}]$ relative to the upper one, which remains fixed (Figure XI.22b). **The fault vector** associated with this stacking fault is thus $-R_1 = 1/6[12\bar{1}]$. The same intrinsic stacking fault could also be obtained with the two other equivalent translations: $-R_2 = 1/6[\bar{1}12]$ and $-R_3 = 1/6[2\bar{1}\bar{1}]$.

- by removing a part of a compact plane (for example, a plane A on figure XI.22c) as a result of vacancy segregations. The resulting fault vector is $R_4 = 1/3[111]$.

Convention
The convention used in electron microscopy is that the lower part of the crystal is shifted with respect to the upper part and not the reverse (Figure XI.21).

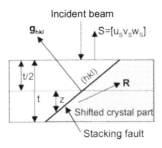

Figure XI.21 - Convention used in electron microscopy for a stacking fault. The lower part of the crystal is shifted with respect to the upper part.

In conclusion, an intrinsic stacking fault can be created by four different mechanisms in a fcc structure. The mechanisms are different but the final result is identical.

Note
The four vectors R_1, R_2, R_3 and R_4 are equivalent since there is always a lattice translation of the type 1/2<110> between any two of them.

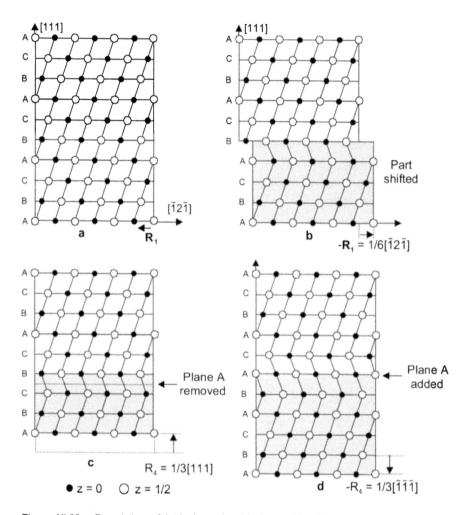

Figure XI.22 - Description of intrinsic and extrinsic stacking faults in face-centred cubic structures.
a - Perfect fcc crystal.
b - Intrinsic stacking fault. The lower part of the crystal is shifted, with respect to the upper part, by a vector $-\mathbf{R_1} = 1/6\,[\bar{1}2\bar{1}]$.
c - Intrinsic stacking fault. A compact atomic plane A is removed. The lower part of the crystal is shifted by a vector $\mathbf{R_4} = 1/3\,[111]$.
d - Extrinsic stacking fault. A compact atomic plane A is added. The lower part of the crystal is shifted by a vector $-\mathbf{R_4} = 1/3\,[\bar{1}\bar{1}\bar{1}]$.

- Extrinsic stacking faults

These faults are obtained by the insertion of a compact atomic plane. This operation results from a concentration of self-interstitials as shown on figure XI.22d where a plane A is added. In this case, the fault vector is $-\mathbf{R_4}=1/3\,[\overline{111}]$.

Finally, intrinsic faults have a fault vector $\mathbf{R} = 1/3{<}111{>}$ and extrinsic faults an opposite fault vector $-\mathbf{R} = 1/3 <\overline{111}>$.

The description given here for compact metal structures can be extended to other types of structures. In the general case, a stacking fault is characterized by an **(hkl) fault plane** and by **a fault vector R** giving the shift between the two parts of the crystal on either side of the stacking fault. The orientation of the fault vector $\mathbf{R}$ with respect to the fault plane depends on the nature of the stacking fault. The full characterization of a stacking fault requires identification of both the (hkl) fault plane and the fault vector $\mathbf{R}$.

XI.3.1.2 - Identification of the fault plane

The (hkl) fault plane can be identified from trace analyses on LACBED patterns according to the procedures and the example described in paragraph X.5.2.

XI.3.1.3 - Identification of the fault vector R

The fault vector $\mathbf{R}$ associated with a stacking fault produces a **phase shift** $\alpha = 2\pi\mathbf{g_{hkl}}.\mathbf{R}$ between the waves emitted by the two parts of the crystal on either side of the fault. In the case of the fcc structure, where the intrinsic and extrinsic stacking faults have a fault vector $\mathbf{R} = \pm1/3{<}111{>}$, the phase shift α can take one of the following three values: $-2\pi/3$, 0 and $+2\pi/3$ ($\pm2k\pi$).

XI.3.1.3.1 - Effect of a stacking fault on diffraction patterns

We have indicated in chapter VII that, within the framework of the two-beam kinematical theory, the intensity I_g diffracted by a perfect crystal is given by the equation:

$$I_g = [1/(s\xi_g)^2]\sin^2(\pi ts) \tag{VII.1}$$

For a crystal containing a stacking fault, the preceding relation becomes: (XI.3)

$$I_g = [\sin^2(\pi ts + \alpha/2) + \sin^2(\alpha/2) - 2\sin(\alpha/2)\sin(\pi ts + \alpha/2)\cos(2\pi sz)] / (s\xi_g)^2$$

In this equation, z defines the height of the stacking fault with respect to the mid-height of the specimen (Figure 21).

When the dynamical theory is used, the corresponding expression given by Hirsch et al. [VI.1] is much more complex:

$$\phi_t = \cos(\pi\Delta kt) - i\,\cos\beta\sin(\pi\Delta kt)$$
$$+ \tfrac{1}{2}\sin^2\beta(e^{i\alpha} - 1)\cos(\pi\Delta kt)$$
$$- \tfrac{1}{2}\sin^2\beta(e^{i\alpha} - 1)\cos(2\pi\Delta kz) \qquad\qquad (XI.4)$$

$$\phi_g = i\,\sin\beta\sin(\pi\Delta kt)$$
$$+ \tfrac{1}{2}\sin\beta(1 - e^{-i\alpha})[\cos\beta\,\cos(\pi\Delta kt) - i\,\sin(\pi\Delta kt)]$$
$$- \tfrac{1}{2}\sin\beta(1 - e^{-i\alpha})[\cos\beta\,\cos(2\pi\Delta kz) - i\,\sin(2\pi\Delta kz)] \qquad (XI.5)$$
$$\text{with } \Delta k = [1 + (s\xi_g)^2]^{1/2}/\xi_g \qquad\qquad (XI.6)$$

In order to take absorption into account, ϕ_t and ϕ_g must be multiplied by $\exp(-\pi t/\xi'_g)$ and Δk must be replaced by:

$$\Delta k = [1 + (s\xi_g)^2]^{1/2}/\xi_g + i/\xi'_g[1 + (s\xi_g)^2]^{1/2} \qquad\qquad (XI.7)$$

The transmitted and diffracted intensities I_t and I_g are thus modified by the presence of a stacking fault and depend both on the phase shift α and on the height z of the fault within the thin foil. Both effects are illustrated on the dynamical rocking curves $I_t = f(s)$ and $I_g = f(s)$ plotted for a stacking fault in a fcc structure in the cases of a small (Figure XI.23) and a large (Figure XI.24) extinction distance.

For a small extinction distance, the transmitted and diffracted rocking curves are strongly modified by the stacking fault (Figure XI.23) and depend on the specimen thickness. Since these effects are not typical they are rather difficult to interpret.

The situation is different for a large extinction distance, where the stacking fault does produce typical effects: the intensity maximum observed on the diffracted rocking curve in the case of a perfect crystal (Figure XI.24a) splits into a principal and a secondary maxima. Depending on the sign of $\alpha = \pm 2\pi/3$, the secondary maximum is located to the left or the right of the principal maximum (Figures XI.24b and d). There is, of course no effect when $\alpha = 0$ (Figure XI.24c). The same behaviour is observed for the transmitted rocking curves (Figure XI.24e, f and g). The most significant effect is obtained when the stacking fault is at z = 0, i.e. at the middle of the specimen (Figure XI.25a). The separation between the principal and the secondary maximum depends on the specimen thickness t (Figures XI.25d, e and f).

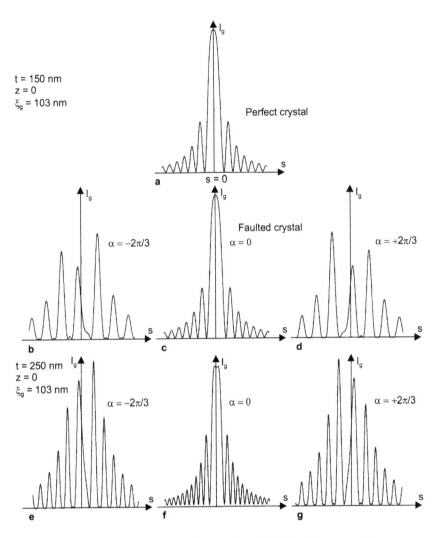

t = 150 nm
z = 0
ξ_g = 103 nm

Perfect crystal

$\alpha = -2\pi/3$

Faulted crystal

$\alpha = 0$

$\alpha = +2\pi/3$

t = 250 nm
z = 0
ξ_g = 103 nm

$\alpha = -2\pi/3$

$\alpha = 0$

$\alpha = +2\pi/3$

Figure XI.23 - Effect of the phase shift α on the diffracted rocking curves for a specimen containing a stacking fault. Dynamical rocking curves in the case of a small extinction distance (example of the 220 silicon reflection with an extinction distance ξ_g = 103 nm at 300 kV). Absorption is not taken into account.
a - Diffracted rocking curve for a perfect crystal.
b, c, d - Diffracted rocking curve for a fcc crystal containing a stacking fault located at the middle of the specimen (at z = 0). The phase shift is α = -2π/3, 0 and +2π/3 and the specimen thickness t is equal to 150 nm.
e, f, g - Same conditions as above except for the specimen thickness is equal to 250 nm.

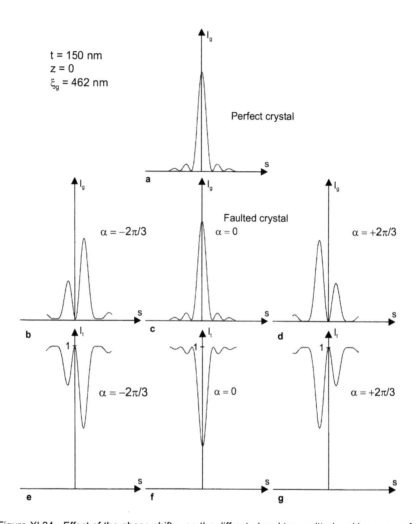

Figure XI.24 - Effect of the phase shift α on the diffracted and transmitted rocking curves for a specimen containing a stacking fault. Dynamical rocking curves in the case of a large extinction distance (example of the 800 silicon reflection with an extinction distance $\xi_g = 462$ nm at 300 kV). Absorption is not taken into account.
a - Diffracted rocking curve for a perfect crystal. It displays a dominant maximum at $s = 0$.
b, c, d - Diffracted rocking curves for a fcc crystal containing a stacking fault located at the middle of the specimen (at $z = 0$). The phase shift is $\alpha = -2\pi/3$, 0 and $+2\pi/3$. A secondary maximum is observed to the right or to the left of the principal maximum, except for $\alpha = 0$.
e, f, g - Transmitted rocking curves corresponding to the curves b, c and d. The same effect is observed.

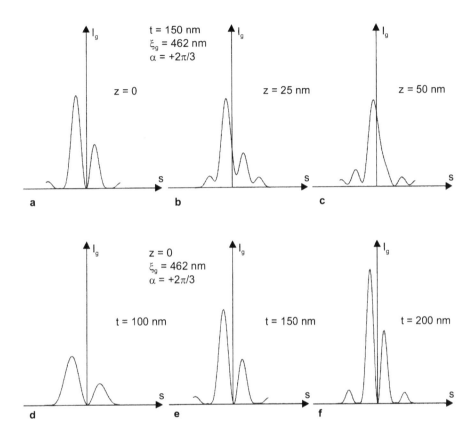

Figure XI.25 - Transmitted rocking curves for a fcc specimen containing a stacking fault with a phase shift $\alpha = +2\pi/3$. Case of a reflection with a large extinction distance.
a, b, c - Effect of the depth z of the stacking fault in the specimen. The best effect is observed for z = 0, i.e. when the stacking fault is in the middle of the specimen.
d, e, f - Effect of the specimen thickness on the separation of the principal and secondary maxima. Note that the specimen thickness does not modify the shape of the rocking curves.

- Effect of absorption

Anomalous absorption does not modify the appearance of the diffracted rocking curves (Figure XI.26a). This is not the case for the transmitted rocking curves, which are strongly modified as shown on figure XI.26b. Quantitative analysis must hence be performed only on the diffracted reflections.

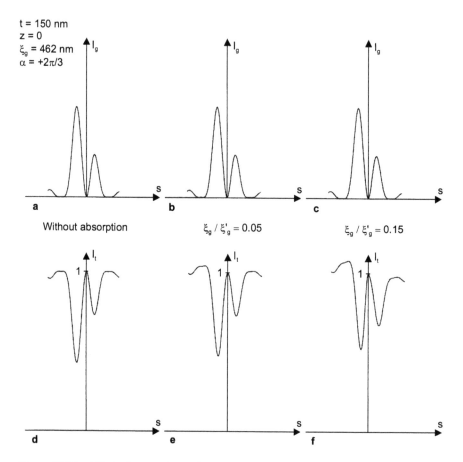

Figure XI.26 - Effect of anomalous absorption on the diffracted and transmitted rocking curves for a specimen containing a stacking fault with a phase shift $\alpha = +2\pi/3$. Case of a reflection with a large extinction distance.
a - Diffracted rocking curve. The shape of the diffracted rocking curve is not modified by absorption.
a - Transmitted rocking curve. The shape of the transmitted rocking curve is modified by absorption.

XI.3.1.3.2 - Effect of a stacking fault on CBED patterns

Although this book is mainly concerned with LACBED, we develop this CBED analysis owing to its importance for stacking fault characterization.

We recall that the diffracted rocking curve is directly observed inside each diffracted disk of a CBED pattern along a diameter perpendicular to the excess line obtained for s = 0 (see paragraph VII.5). In the same way, all the transmitted rocking curves are observed inside a single transmitted disc.

When the incident beam is focused on a tilted stacking fault (if possible at the middle of the fault), we can expect that each excess and deficiency line whose dot product $g_{hkl}.R$ is not equal to an integer will split into a principal and a secondary line. The CBED pattern on figure XI.27b obtained from a silicon specimen observed along the [519 476 1701] direction shows that this is indeed the case.

- Identification of the fault vector R from CBED patterns

The identification of the fault vector requires knowledge of:
- the hkl indices of the excess or deficiency lines present on the pattern,
- the positive direction of the deviation parameter s.
- the position of the secondary line with respect to the principal one.

The sign of α can be obtained from the rules given on figure 28. These rules are directly deduced from the rocking curves on figure XI.23.

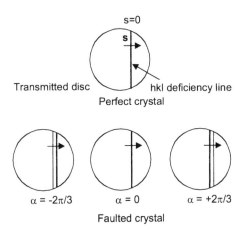

Figure XI.28 - Rules used to identify the phase shift α from the splitting of the deficiency line in the case of a stacking fault present in a fcc structure.
The position of the secondary line to the right or to the left of the principal line allows the phase shift α to be identified provided that the positive direction of the deviation parameter s is known.

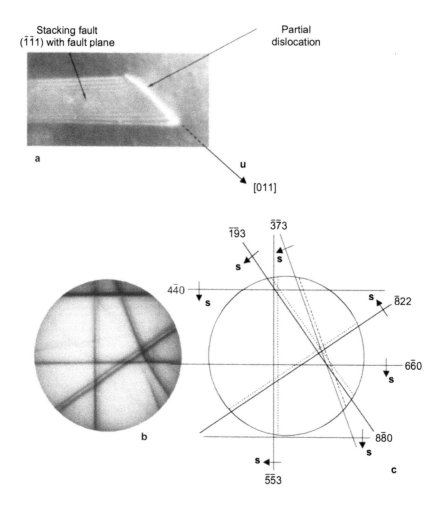

Figure XI.27 - Effect of a stacking fault with a ($\bar{1}\bar{1}1$) fault plane on a bright field CBED pattern.

a - Micrograph of the stacking fault.

b - Experimental CBED pattern observed along the [519 476 1701] specimen orientation. Most of the deficiency lines are split into a principal and a secondary line.

c - Theoretical CBED pattern from a specimen containing a stacking fault with a fault vector

R = 1/3 [11$\bar{1}$] . It is in agreement with the experimental pattern.

The partial dislocation bordering this stacking fault will be characterized in the paragraph XI.3.2.2.

Simulation carried out with the "Electron Diffraction" software [VI.3]

Silicon specimen. Courtesy of J.W. Steeds

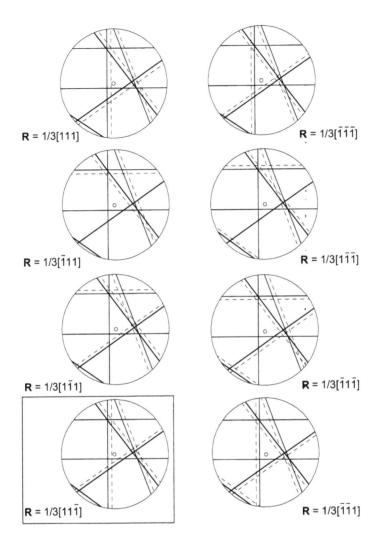

Figure XI.29 - Theoretical bright field CBED patterns drawn for the eight possible fault vectors in the case of a silicon specimen observed along the [519 476 1701] direction. The secondary lines are represented by dotted lines. All patterns are different, meaning that the fault vector **R** can be identified without ambiguity from this specimen orientation. The pattern for **R** = 1/3 [11$\bar{1}$] is in agreement with the experimental pattern on figure XI.27b.

Simulations carried out with the software "Electron diffraction" [VI.3].

In order to characterize the fault vector **R**, a first solution consists in comparing the position of the secondary lines on the experimental pattern with their position on theoretical patterns established for all the possible fault vector. On these simulations, the secondary lines are drawn with dotted lines and their location with respect to the principal lines is deduced from the rules on figure XI.28. In the case of fcc structures, eight possible fault vectors of the type 1/3<111> can be encountered. The corresponding simulated patterns are given on figure XI.28. All of them are different, meaning that there is no ambiguity in the identification of **R** from these patterns. Using this procedure, the fault vector **R** = 1/3[11$\bar{1}$] is identified for the experimental pattern on figure XI.27a by straightforward comparison with the theoretical patterns on figure XI.28.

A second possibility consists in comparing the experimental α values with those calculated for all the possible fault vectors as shown on table XI.3. The fault vector **R** = 1/3[11$\bar{1}$] is also easily identified from this table.

g_{hkl}	α_{exp}	R = ±1/3 [111]	R = ±1/3 [$\bar{1}$11]	R = ±1/3 [1$\bar{1}$1]	R = ±1/ [11$\bar{1}$]
$\bar{1}93$	-2π/3	∀2π/3	±2π/3	∀2π/3	∀2π/3
$\bar{8}22$	+2π/3	∀2π/3	0	±2π/3	±2π/3
$\bar{3}\bar{7}3$	-2π/3	∀2π/3	∀2π/3	±2π/3	∀2π/3
$\bar{5}\bar{5}3$	-2π/3	∀2π/3	0	0	∀2π/3
$8\bar{8}0$	0	0	±2π/3	±2π/3	0
$6\bar{6}0$	0	0	0	0	0
$4\bar{4}0$	0	0	±2π/3	∀2π/3	0
		No	No	No	Yes

Table XI.3 - Experimental values of the phase shifts α obtained from the experimental pattern on figure XI.26. The values of the phase shifts for the eight possible fault vectors R are also given in this table. The fault vector **R** = 1/3[11$\bar{1}$] is easily identified from this table.

XI.3.1.3 3 - Effect of a stacking fault on LACBED patterns

As shows on figure XI.30a, the effect of a stacking fault is more complex on LACBED patterns than on CBED patterns. For a tilted stacking fault perpendicular to a Bragg line, z varies according to the Bragg line. In order to simulate the effect of a stacking fault, the rocking curves must be drawn for all the z values contained between the two limiting values ± t/2.

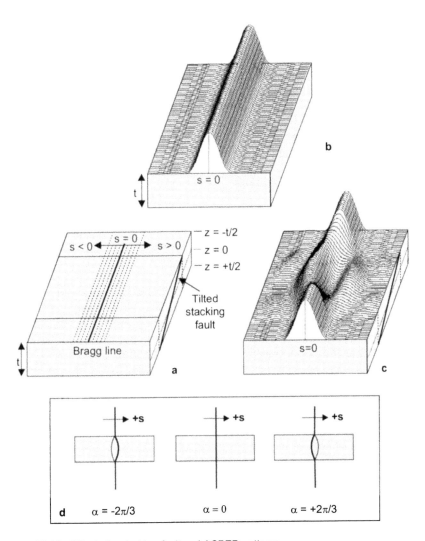

Figure XI.30 - Effect of a stacking fault on LACBED patterns.
a - Principle of calculation of the effect of a stacking fault on a Bragg line for a tilted stacking fault perpendicular to the Bragg line.
b - Simulation of a Bragg line for a perfect crystal.
c - Simulation of a Bragg line for a crystal containing a tilted stacking fault with $\alpha=+2\pi/3$. The Bragg line is split into a principal and a secondary line.
d - Identification rules for the phase shift from observation of the Bragg line splitting. The positive direction of the deviation parameter s must be taken into account.

Dynamical simulations (figure XI.30c) show that the Bragg line splits into a principal and a secondary line when crossing the stacking fault. The maximum separation of the two lines is observed at the middle of the fault, i.e. for $z = 0$.

The secondary line is located either on the right or on the left of the principal line, depending on the sign of the phase shift α and the positive direction of the deviation parameter s. The phase shift can hence be identified from the rules given in figure XI.30d and the fault vector **R** is deduced by means of the procedures described for CBED patterns.

The experimental patterns on figure XI.31 show the effect of a stacking fault on some excess and deficiency Bragg lines.

XI.1.1.3.4 - Operating conditions

For CBED patterns, it is advisable to use a very small probe size and to position it as closely as possible in the middle of the stacking fault in order to obtain the best splitting between the principal and the secondary line.

For LACBED patterns, the best effects are obtained when the stacking fault is tilted inside the thin foil and when the Bragg line forms an angle close to 90° with the trace of the stacking fault (Figures XI.32a, b and c). If this angle is different from 90°, the separation between the principal and the secondary lines becomes distorted (Figure XI.32c). For the special case of a stacking fault trace parallel to the Bragg line, the splitting consists of two parallel principal and secondary lines, the position of the secondary line relative to the principal line being determined by the phase shift α and the deviation parameter s (Figure XI.32d).

The effects of the Δh value as well as its sign are also illustrated on figures XI.32a and b. We note that the position of the secondary line with respect to the principal one does not depend on the sign of Δh (Figures XI.32 and 33).

The analysis given above applies to quasi kinematical Bragg lines, i.e. to lines with relatively large extinction distances ξ_g. It cannot be applied to dynamical Bragg lines consisting of a set of fringes. In this case, the stacking faults produce very complex effects depending dramatically on the specimen thickness. Two examples are given on figures XI.31a and c. Since these effects are not typical, they are used for the identification of the fault vector **R** only in some special cases.

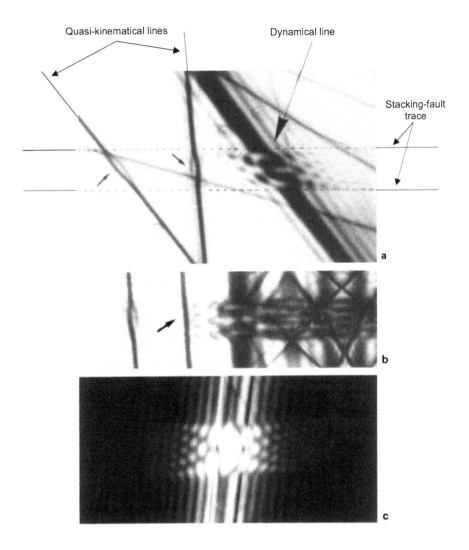

Figure XI.31 - Effect of a stacking fault on some Bragg lines present on bright- and dark-field LACBED patterns.
a - Bright-field LACBED pattern. The stacking fault intersects two quasi-kinematical lines and a dynamical line. For the latter, the appearance is very complex.
b - Bright-field LACBED pattern. The stacking fault intersects three Bragg lines with multiple orders. There is no effect for the Bragg line marked by an arrow because the dot product $g_{hkl}.R$ is an integer.
c - Dark-field LACBED pattern. The stacking fault intersects a dynamical line producing a very complex pattern.

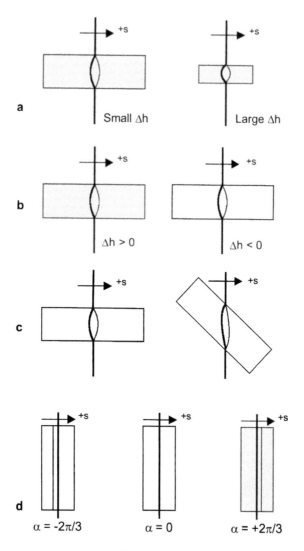

Figure XI.32 - Schematic description of the effect of a stacking fault on Bragg lines present on bright-field LACBED patterns.

a - Effect of the Δh value. The image magnification is modified.

b - Effect of the Δh sign. The position of the secondary line is not altered.

c - Effect of the orientation of the fault trace with respect to the Bragg line. The separation of the principal and secondary lines is distorted.

d - Particular case observed when the fault trace is parallel to the Bragg line. The secondary line is located to the right or to the left of the principal line depending on the value of the phase shift α and on the sign of the deviation parameter s.

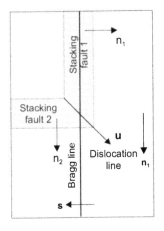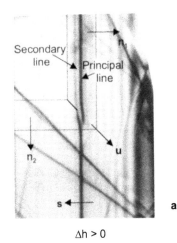

Δh > 0

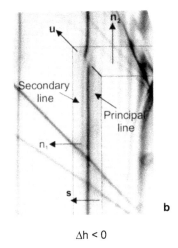

Δh < 0

Figure XI.33 - Effect of the specimen height Δh on the relative position of the secondary line with respect to the principal line on bright-field LACBED patterns.
a - Bright-field LACBED pattern containing two stacking faults and a partial dislocation line. For Δh > 0, the secondary line for the stacking fault 1 is located to the left of the principal line.
b - For Δh < 0, the image is rotated by 180° but the secondary line remains to the left of the principal one. Note that the dislocation splitting is mirror-related to the splitting on figure a. Silicon specimen. Courtesy of P. H. Albarède.

XI.3.1.3.5 - Quantitative analysis

The foregoing analysis is qualitative. Quantitative analysis can also be performed to measure unknown phase shifts. As shown on figure XI.34, the relative intensity of the main and secondary maxima of the diffracted rocking curves depends on the phase shift. These phase shifts can be obtained from accurate measurement of the transmitted or diffracted intensities across a splitting with a CCD camera or with "Imaging Plates". Observations of dark-field patterns are recommended, since they are not affected by anomalous absorption. Some examples are given in reference [XI.5].

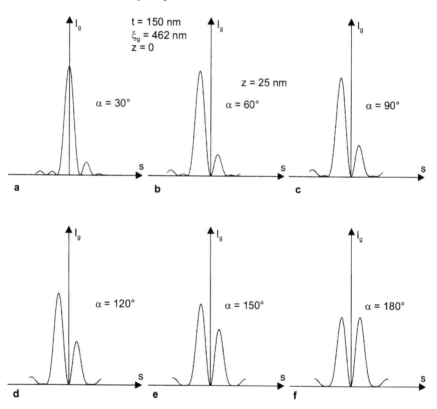

Figure XI.34 - Effect of the phase shift on the diffracted rocking curves in the case of a reflection with a large extinction distance. The relative intensity of the principal and secondary maxima depends on the phase shift and can be used to identify it.

XI.3.1.4 - Advantages of the CBED and LACBED methods with respect to the conventional techniques

In conventional electron microscopy, the fault vector **R** is generally identified by looking for specific crystal orientations that give non-contrasted micrographs of the defects. In this case, the phase shift α equals zero or $2k\pi$ and the dot product $\mathbf{g_{hkl}.R}$ equals 0 or an integer. At least two experiments are required to obtain **R**.

The CBED analysis is much easier to perform since a single CBED pattern without a specific specimen orientation is usually all that is required. In addition, the simultaneous observation of several splittings on the same pattern gives a reliable result. The main difficulty concerns the specimen contamination associated with the small spot sizes used for the analyses.
The LACBED analysis can also be used but it requires observation of at least three splittings.

XI.3.2 - Characterization of partial dislocations

XI.3.2.1 - Crystallography of partial dislocations

A stacking fault can end at the external surface of the crystal or inside it. In this second case, it is bordered by two partial dislocations (Figure XI.35) or by a dislocation loop.

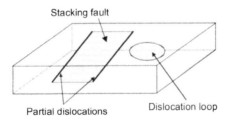

Figure XI.35 - A stacking fault located inside a crystal is limited by two partial dislocation lines or by a dislocation loop.

We reconsider fcc structures. We can expect that dislocations in these structures will have a Burgers vector of the type **b** = 1/2<110> corresponding to the smallest lattice translations (Figure XI.36a). This is indeed what is observed. We examine the example of an edge dislocation with a Burgers vector **b** = 1/2 [$\bar{1}$10] located in a (111) slip plane

(Figure XI.36b). This plane contains the dislocation line **u** and the Burgers vector **b**. As show on figure XI.36c, the additional atomic plane is a $(1\bar{1}0)$ plane and the dislocation line is along the $[\bar{1}\bar{1}2]$ direction. If we project the trace of the $(1\bar{1}0)$ planes along the [111] direction of the fcc structure (Figure XI.36d), we note that these planes are stacked according to a sequence ababab...and that the Burgers vector **b** corresponds to two $(1\bar{1}0)$ atomic planes. Thus, the distorted zone, surrounding the dislocation line, results from the insertion of two $(1\bar{1}0)$ atomic planes, as shown on figure 36e.

Such a dislocation moves by the displacement of the (a+b) unit, i.e. by the displacement of two $(1\bar{1}0)$ atomic planes. Note that this displacement does not modify the stacking sequence of the (111) planes (the planes drawn with dotted or continuous lines are in register above and below the (111) glide plane). Nevertheless, this movement is not energetically favourable. To understand this point, consider an atom at a position B in a layer located above the slip plane where all the atoms occupy positions A (Figure XI.37a). The displacement of this atom from a position B to another position B corresponds to the Burgers vector **b**. It forces this atom to pass above the atom A_1. A better solution is to follow the valley between the two A_1 and A_2 atoms in order to reach a C position, and then to continue the travel by following the valley between the A_1 and A_3 atoms finally to reach a position B. This "zigzag" travel produces the same result as the previous direct travel but requires less energy. The displacement from B to C corresponds to a dislocation with Burgers vector b_1 and to the first additional $(1\bar{1}0)$ atomic plane. The second displacement from C to B corresponds to second dislocation with a vector b_2 and to the second additional atomic plane.

As a result:

$b_1 + b_2 = b$

b_1 and b_2 are the two partial dislocations with:

$b_1 = 1/6\,[\bar{2}11]$ and $b_2 = 1/6\,[1\bar{2}\bar{1}]$.

The perfect dislocation with vector $b = 1/2\,[\bar{1}10]$ is dissociated into two partial dislocations b_1 and b_2 according to the dissociation reaction:

$\frac{1}{2}[\bar{1}10] \ \tau \ 1/6\,[\bar{2}11] + 1/6\,[1\bar{2}\bar{1}]$ (Figure XI.37b). The introduction of the first atomic plane moves all the atoms located above the slip plane by a vector b_1. The second atomic plane moves them by b_2.

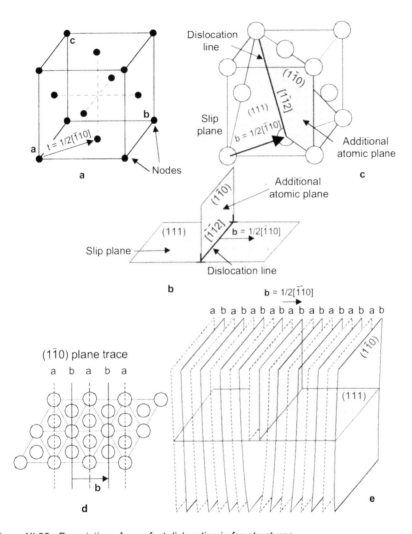

Figure XI.36 - Description of a perfect dislocation in fcc structures.

a - Description of the direct lattice of the fcc structure. The smallest lattice translations are of the type 1/2<110>.

b - Perfect edge dislocation with a Burgers vector **b** = 1/2 [$\bar{1}$10] and a slip plane (111).

c - Description of the fcc structure showing the [$\bar{1}\bar{1}$2] dislocation line and the ($\bar{1}$10) additional atomic plane.

d - Projection of the ($\bar{1}$10) lattice planes along the [111] direction. The stacking sequence of these planes is ababab..... The Burgers vector **b** corresponds to the (a+b) unit.

e - Perfect edge dislocation. The distorted zone around the line is formed by the insertion of two additional ($\bar{1}$10) atomic planes. The stacking sequence of the (111) lattice planes is not modified (the dotted and continuous lines are in register above and below the slip plane).

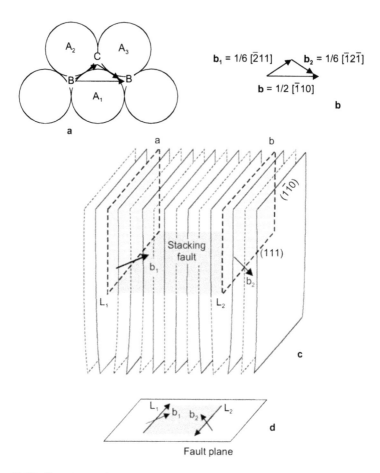

Figure XI.37 - Description of a partial dislocation in fcc structures.

a, b - The shift from a position B to another position B corresponds to the Burgers vector **b** of a perfect dislocation. It is energetically unfavourable. The shift from B to C followed by a second shift from C to B is more favourable. It corresponds to two partial dislocations with Burgers vectors **b₁** and **b₂**.

c - The partial dislocation line L₁ moves all the atoms located above the slip plane by a vector **b₁**. The second line L₂ moves them by a vector **b₂**. A stacking fault ribbon is created between the L₁ and L₂ partial dislocation lines (the $(1\bar{1}0)$ planes drawn with dotted and continuous lines are not in register above and below the glide plane).

d - The L₁ and L₂ dislocation lines and the **b₁** and **b₂** Burgers vectors are located in the stacking-fault plane. These partial dislocations are called glissile partial Shockley dislocations.

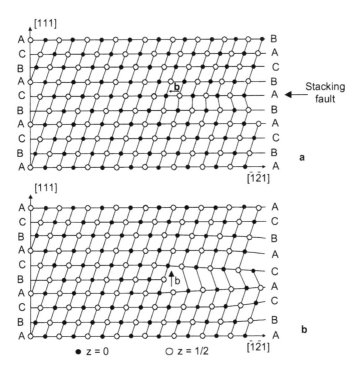

Figure XI.38 - Description of the distorted zones located around partial dislocation lines in fcc structures.
a - Shockley partial dislocation. The Burgers vector **b** is contained in the stacking-fault plane.
b - Frank partial dislocation. The Burgers vector b is perpendicular to the stacking-fault plane.

If the two ($1\bar{1}0$) planes separate, as is the case on figure XI.37c, then a stacking fault ribbon is created between them (the ($1\bar{1}0$) planes drawn with dotted and continuous lines are not in register above and below the glide plane).

In the present case, the two dislocation lines, the two Burgers vectors and the stacking fault all are located in the same plane (Figure XI.37d). Since a dislocation is forced to slip in a plane containing both the dislocation line and its Burgers vector, this means that the two partial dislocations can slip in the fault plane. For this reason, they are called **glissile Shockley partial dislocations** (Figure XI.38a).

We note that the Burgers vectors of the partial dislocations are of the same type as the fault vectors R_1, R_2 and R_3.

For fcc structures, there is a second type of partial dislocation bordering a stacking fault. Imagine that a part of the B plane on figure XI.38b disappears under the action of vacancy clustering. An intrinsic stacking fault is created and it is bordered by two **Frank partial dislocations** with Burgers vector $b = 1/3<111>$. These partial dislocations cannot slip because their slip plane does not coincide with the stacking fault plane. They are called **sessile partial dislocations**. In the same way, a local segregation of the self-interstitial atoms creates an extrinsic stacking fault bordered by two Frank partial dislocations with opposite Burgers vector $b = 1/3 <\overline{111}>$.

XI.3.2.1.1 - Burgers circuit of partial dislocations

The Burgers circuit of a partial dislocation must be drawn with the FS/RH conventions. Moreover, the circuit must begin and finish on the fault plane (Figure XI.39).

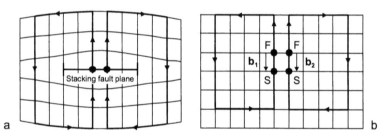

a

b

Figure XI.39 - Burgers circuit for a partial dislocation.
The FS/RH circuit must start and end on the stacking fault. The vector **u** of the right partial dislocation points into the page and that of the left dislocation points out of the page.
a - Burgers circuit in the faulted crystal.
b - Burgers circuit in the perfect crystal.

XI.3.2.2 - Identification of partial dislocations from LACBED patterns

The three examples of Burgers vector identification given in paragraphs XI.1.2.1.1, XI.1.2.1.2 and XI.1.2.1.3 relate to perfect dislocations. In this case, the dot product:
$g_{hkl}.b = n = (ha^* + kb^* + lc^*).(u_b a + v_b b + w_b c) = hu_b + kv_b + lw_b$ is always an integer since the Miller indices h, k, l are, by definition, integers and the Burgers vector $b = [u_B v_B w_B]$ is a lattice translation. With partial dislocations this is not always true.

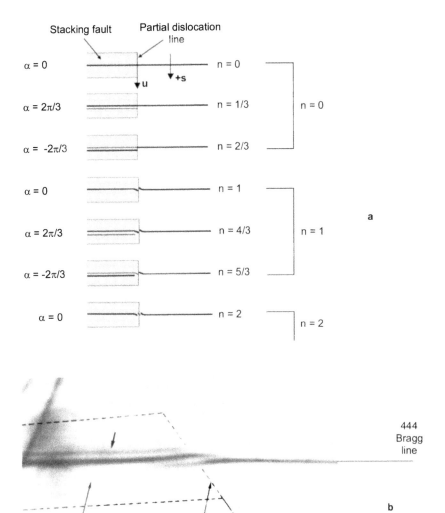

Figure XI.40 - Effect of Shockley and Frank partial dislocations on Bragg lines of LACBED patterns.
a - Simulations of the intersection of a partial dislocation with a Bragg line. The splittings for n + 1/3 and n + 2/3 are the same as those observed for n in the case of a perfect dislocation. Simulations according to Tanaka [XI.5].
b - LACBED pattern showing the splitting of a Frank partial dislocation on the 444 Bragg line with n = - 4/3. Note the splitting of the Bragg line in the stacking fault.
Silicon specimen. Courtesy of J. W Steeds.

We have mentioned that the partial dislocations most frequently observed in fcc structures have Burgers vectors of the type $\mathbf{b} = 1/6<112>$ for Shockley partial dislocations and $\mathbf{b} = 1/3<111>$ for Frank partial dislocations. They are associated with stacking faults with fault vector $\mathbf{R} = \pm<111>$. For these partial dislocations, the possible values of the dot product $\mathbf{g}.\mathbf{b} = n$ are 0, $\pm1/3$, $\pm2/3$, ±1, $\pm4/3$, $\pm5/3$, ±2, $\pm7/3$..... These values are not all integers. Simulations carried out by Tanaka et al. [XI.5] show that the splittings for $n + 1/3$ and $n + 2/3$ look like those observed for n in the case of a perfect dislocation (Figure XI.37a). It becomes thus difficult to identify the true value of n.

How can this problem be solved?

The simplest solution consists in looking for the experimental conditions that give an integer dot product $\mathbf{g}_{hkl}.\mathbf{b} = n$. These conditions are obtained when the dot product $\mathbf{g}_{hkl}.\mathbf{R}$ for the associated stacking fault is equal to zero or to an integer. Experimentally, these conditions are easy to find since they correspond to the absence of effect of the stacking fault on the Bragg line involved. To characterize partial dislocations, therefore, we consider only the splittings with Bragg lines unaffected by the stacking fault. This solution was adopted for the characterization of the partial dislocation located at the end of the stacking fault shown on figure XI.27a. Its fault vector $\mathbf{R} = 1/3\,[11\bar{1}]$ was already identified in chapter XI.3.1.3.2. We have considered the intersections of this partial dislocation with the three Bragg lines $\bar{6}\bar{6}0$, $0\bar{8}4$ and $010\bar{2}$ (the three corresponding $\mathbf{g}_{hkl}$ vectors are not coplanar) whose dot product $\mathbf{g}_{hkl}.\mathbf{R}$ is an integer (Figure XI.38). The corresponding values of the splitting numbers n are given in table XI.4.

$\mathbf{g}_{hkl}$	n	$R = 1/3\,[11\bar{1}]$	$b = \pm1/3\,[\bar{1}\bar{1}1]$	$b_I = \pm1/6\,[2\bar{1}1]$	$b_{II} = \pm1/6\,[1\bar{2}1]$	$b_{III} = \pm1/6\,[112]$
$\bar{6}\bar{6}0$	+4	+4	±4	∀1	±1	∀2
$0\bar{8}4$	-4	+4	∀4	∀2	∀2	0
$010\bar{2}$	-4	+4	∀4	∀2	∀3	±1
			Yes	No	No	No

Table XI.4 - Experimental values of the splitting number n obtained from figure XI.38. The values of the dot products $\mathbf{g}_{hkl}.\mathbf{R}$ for the possible Frank and Shockley partial dislocations are also indicated in this table. From this table the Burgers vector is seen to be $\mathbf{b} = 1/3\,[\bar{1}\bar{1}1]$.

In this table, we also indicate the values of the $\mathbf{g}_{hkl}.\mathbf{b}$ products for all the possible Frank and Shockey partial dislocations associated with a stacking fault with $\mathbf{R}=1/3\,[11\bar{1}]$. For Shockley dislocations, the Burgers vectors of the type 1/6<112> must lie within the fault plane. The vectors $\mathbf{b_I} = \pm 1/6\,[2\bar{1}\bar{1}]$, $\mathbf{b_{II}} = \pm 1/6\,[1\bar{2}\bar{1}]$ and $\mathbf{b_{III}} = \pm 1/6\,[112]$ are in agreement with this condition.

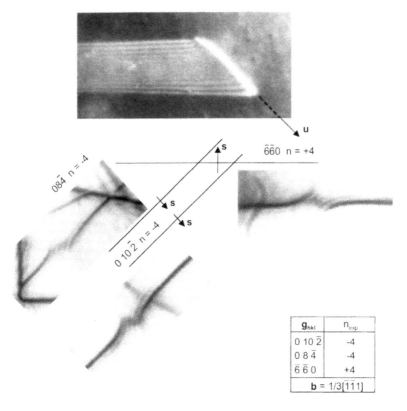

g_{hkl}	n_{exp}
0 10 $\bar{2}$	-4
0 8 $\bar{4}$	-4
$\bar{6}\,\bar{6}\,0$	+4
$\mathbf{b} = 1/3[\bar{1}\bar{1}1]$	

Figure XI.41 - Identification of the Burgers vector of a Frank partial dislocation located at the end of the stacking fault studied in paragraph XI.3.1.3.2 and on figure XI.27.
LACBED patterns showing the splittings of the partial dislocation line with the Bragg lines $\bar{6}\bar{6}0$, $08\bar{4}$ and $010\bar{2}$.
The Bragg lines were selected in order to give integer values of n. They do not produce any effect on the stacking fault. The Burgers vector is $\mathbf{b} = 1/3\,[\bar{1}\bar{1}1]$.
Silicon specimen. Courtesy of J.W. Steeds.

These results lead to a Burgers vector $\mathbf{b} = 1/3\,[\bar{1}\bar{1}1]$. This vector is perpendicular to the stacking fault and the dislocation line $\mathbf{u}$ is along the [011] direction. The dislocation is, therefore, an edge Frank partial dislocation.

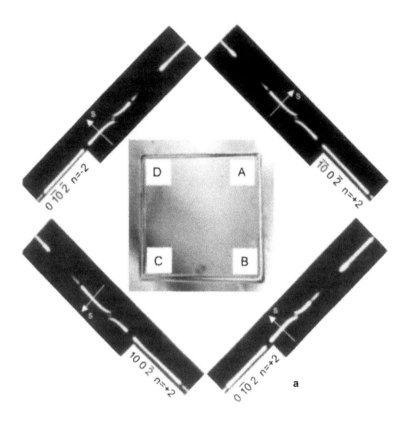

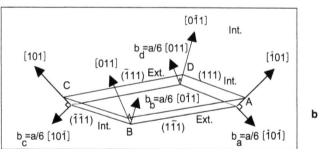

Figure XI.42 - Characterization of the Burgers vector of four Lomer-Cottrell partial dislocations located at the intersections of a stacking fault pyramid.
a - Dark-field LACBED patterns. Bragg lines of the type {10 0 2} are selected because they are not affected by the stacking faults.
b - Result of the LACBED analysis.
Silicon specimen. Courtesy of P. H. Albarède

Note that the n values are different in each of the columns of table XI.4, which means that there is no ambiguity in the identification of the Burgers vector of partial dislocations from these experimental conditions.

The second example concerns partial dislocations present at the intersection of four stacking faults forming a regular pyramid (Figure XI.42a). In this case, the experimental conditions are more complex than before since the two stacking faults bordering each partial dislocation must not affect the Bragg lines used for the analysis. For example, in the case of the partial dislocation located at the intersection of the (111) and ($1\bar{1}1$) fault planes (point A on figure XI.42), we have selected the $\overline{10}0\,\overline{2}$ line because it satisfies the previous condition $\mathbf{g}\,_{\overline{10}0\,\overline{2}}.1/3\,[111] = -4$ and $\mathbf{g}\,_{\overline{10}0\,\overline{2}}.1/3\,[1\bar{1}1] = -4)$.

The complete analysis of this stacking fault pyramid [XI.11] leads to Burgers vectors of the type 1/6<110> (Figure XI.39b). They correspond to Lomer-Cottrel partial dislocations (stair-rods dislocations). These dislocations are generally very difficult to characterize with conventional methods owing to the low value of their Burgers vector.

In the two preceding examples, we have chosen experimental conditions giving integer n values and the Burgers vector is identified directly by visual observation of the splittings.

How can we identify non-integer n values?

A solution consists in observing accurately the appearance of the Bragg line at the intersection with the partial dislocations and also inside the associated stacking fault. [XI.5]. Comparison with the simulations on figure XI.37a allows non-integer n values to be identified. This type of analysis is, however, very difficult to carry out. It requires that the Bragg line implied in the analysis be nearly parallel to the stacking fault.

XI.3.2.3 - Characterization of dislocation loops

XI.3.2.3.1 - Crystallography of dislocation loops

Perfect or partial dislocation lines can also form closed loops, which adopt various configurations according to the orientation of the Burgers vector **b** with respect to the loop.

If the vector **b** is parallel to the dislocation loop, then the loop is described as a **slip loop** (Figure XI.40a). Depending on its location along the line, the loop has a pure edge, screw or mixed character. A Shockley partial dislocation loop surrounding an intrinsic stacking fault belongs to this category.

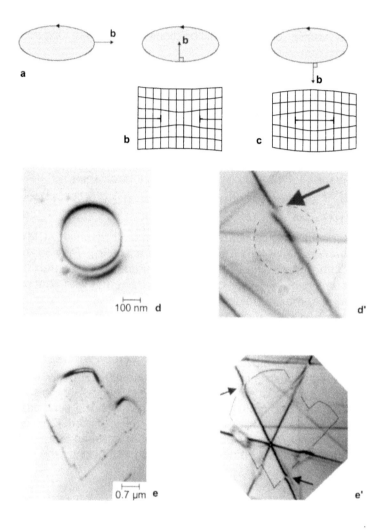

Figure XI.43 - Schematic description of dislocation loops.
a - Glide loop. The vector **b** is contained in the fault plane.
b, c - Interstitial and vacancy loops. The vector **b** is perpendicular to the fault plane.
d, d' - Bright-field image of a small vacancy loop and corresponding LACBED pattern.
e, e' - Bright-field image of an interstitial loop and corresponding LACBED pattern.
GaP specimen and patterns. Courtesy of W. Jäger and C. Jäger.

If the vector **b** is not located in the plane of the dislocation loop, the loop is known as a **prismatic loop**. If the vector **b** is perpendicular to the loop, the dislocation has an edge character everywhere along the line. A Frank partial dislocation surrounding an intrinsic or an extrinsic stacking fault produces **an interstitial or a vacancy loop** (Figures XI.40b and c).

XI.3.2.3.2 - Identification of dislocation loops from LACBED patterns

Dislocation loops can be identified using the same procedures as those described for perfect or partial dislocations. In the case of large loops, the analysis is straightforward: when a Bragg line intersects the loop, it produces two splittings whose quality depends on the location of the intersection (Figures XI.43d' and e'). We recall that the best splittings are obtained when the Bragg line intersects the part of the line that is located in the middle of the specimen. For a partial dislocation loop, the intrinsic or extrinsic stacking fault located inside the loop also produces an effect.

Many difficulties are encountered with small loops because the effects of the loop and that of the stacking fault are more or less superimposed. Identification rules are given in reference [XI.4].

XI.3.3 - Characterization of antiphase boundaries (APB)

XI.3.3 1 - Crystallography of antiphase boundary

Many metallic solid solutions are disordered at high temperature and ordered at low temperature. Consider the example of the β phase of the Cu-Al-Ni shape-memory alloys [XI.15]. Depending on the temperature, the Cu, Al and Ni atoms are differently disposed on the two atomic sites of a bcc structure, i.e. at 000 (site α) and at ½ ½ ½ (site β) (Figure XI.44a).

At high temperature, the atoms are randomly disposed on these two sites and the β phase is disordered. To describe it, we can consider that each α and β site is statistically occupied by an average atom composed of a mixture of Cu, Al and Ni, taking into account the average composition (Figure XI.41c). The resulting structure is a body-centred cubic structure belonging to the Im$\bar{3}$m space group.

At low temperature, all the Al atoms and a part of the Cu atoms are located on the α sites. The β sites are occupied by all the Ni atoms and the remainder of the Cu atoms. This ordered β_2 phase is primitive cubic and belongs to the space group $P\bar{4}3m$.

The corresponding diffraction patterns contain two types of reflections: fundamental and superlattice reflections. The fundamental reflections are common to both the ordered and disordered phases and are observed when the sum $h + k + l$ of the Miller indices is odd. Superlattice reflections are present in the ordered phase only when the sum $h + k + l$ is even. Superlattice reflections are therefore typical of the ordering and can be used to identify it. The structure factor is $F_{hkl} = f_\alpha + f_\beta$ for the fundamental reflections and $F_{hkl} = f_\alpha - f_\beta$ for the superlattice reflections, f_α and f_β being the atomic scattering factors associated with the α and β sites. This means that the superlattice reflections are much weaker than the fundamental ones.

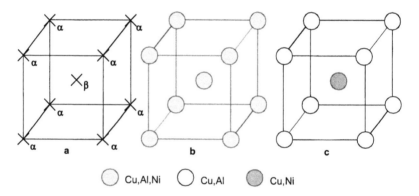

Figure XI.44 - Description of the ordered β_2 and disordered β phases of the Cu-Al-Ni shape-memory alloys.
a - Description of the atomic sites α at 000 and β at ½ ½ ½ of the bcc structure.
b - Description of the disordered β phase. The Cu, Al and Ni atoms are randomly located on the α and β sites. These sites are statistically occupied by an average atom composed of a mixture of Cu, Al and Ni atoms taking into account the average alloy composition.
b - Description of the ordered β_2 phase. All the Al atoms and some of the Cu atoms are located on the α sites. The other atoms are located on the β sites.

Upon cooling from high temperature, the disordered β phase undergoes an ordering transformation (B2 ordering) by nucleation and growth. The β_2 ordered nuclei develop and can either be "in phase" or "in antiphase" with respect to each other when they unite (Figure XI.45). In the second case, a crystal defect called an **antiphase boundary (APB)** is created at the interface between the two domains. One domain is shifted with respect to the other one by any of the eight equivalent vectors $R = \frac{1}{2}<111>$, i.e. by a translation of the disordered structure (but not of the ordered one). This antiphase boundary can also be considered as a stacking fault in an ordered structure.

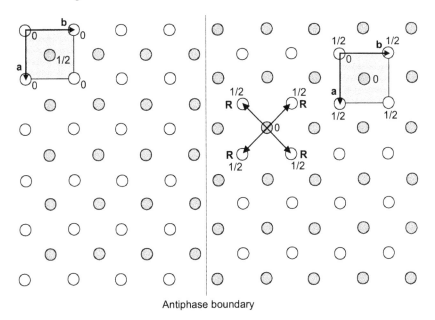

Antiphase boundary

Figure XI.45 - Schematic description of an antiphase boundary (APB).
The right part of the ordered crystal is shifted with respect to the left part by any of the eight equivalent vectors $R = \frac{1}{2}<111>$.

XI.3.3.2 - Effect of APB on diffraction patterns

Since an antiphase boundary can be regarded as a special type of stacking fault, we can expect to observe effects on the diffraction patterns similar to those observed with stacking faults, i.e. splittings of some reflections. For the present antiphase boundaries, the fault vectors R are of the type $\frac{1}{2}<111>$ so that the dot products $g_{hkl}.R$ have integer values for the fundamental reflections and values $\pm 1/2, \pm 3/2, \pm 5/2, \pm 7/2...$ for the superlattice reflections. The corresponding phase shifts $\alpha = 2\pi g_{hkl}.R$ are equal to 0 and $\pm \pi$. For this reason APBs are often called π defects.

As a consequence, only the superlattice reflections will be affected by APBs. These reflections are usually weak and have a large extinction distance so that only the strongest ones can be observed easily. They correspond to low hkl indices and to small Bragg angles.

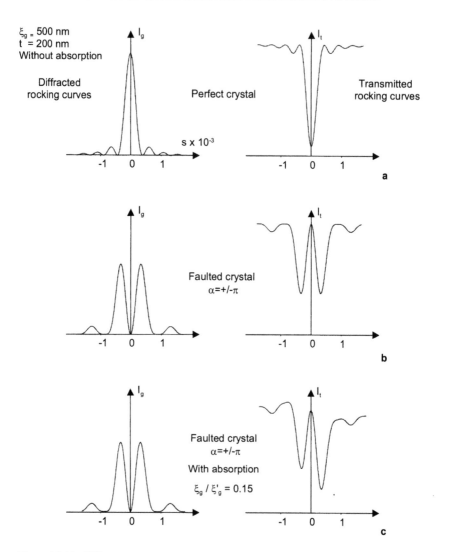

Figure XI.46 - Diffracted and transmitted dynamic rocking curves for a crystal containing an antiphase boundary. The calculations are made for a superlattice reflection with a large extinction distance.

a - Diffracted and transmitted dynamic rocking curves for a perfect crystal.

a - Diffracted and transmitted dynamic rocking curves for a crystal containing an APB. The rocking curves are split into two maxima or minima having the same intensity.

b - Diffracted and transmitted dynamic rocking curves for a crystal containing an APB when anomalous absorption is taken into account. The diffracted rocking curve remains symmetrical with respect to s = 0 but the transmitted one is no longer symmetrical.

ξ_g= 500 nm
t = 200 nm
With absorption

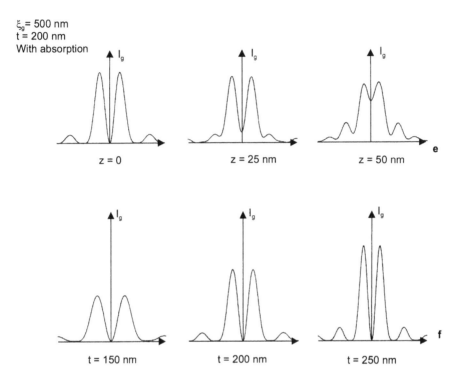

z = 0   z = 25 nm   z = 50 nm   e

t = 150 nm   t = 200 nm   t = 250 nm   f

Figure XI.46 - Continuation.
e - Effect of the depth z of the APB in the specimen. The best effect is observed when z = 0, i.e. when the APB is in the middle of the specimen.
f - Effect of the specimen thickness on the separation of the two splitting maxima.

This means that APB effects will be observed on reflections having at the same time a large extinction distance and a small Bragg angle. A few reflections are in agreement with these conditions. We recall that, for a stacking fault, all the reflections whose dot product $g_{hkl}.R$ is non integer are affected by a stacking fault.

The expected effects can be obtained from the dynamical rocking curves drawn from equations (XI.4 and 5) and shown on figure XI.46. These curves indicates that the diffracted and transmitted rocking curves split symmetrically into two maxima with equal intensity (Figure XI.46b). If absorption is taken into account (Figures XI.46c and d), the intensity of the two maxima remains equal on the diffracted rocking curves but not on the transmitted curves, which are no longer symmetrical with respect to s = 0.

The best effect is observed when the APB is located in the middle of the specimen (at $z = 0$) (Figure XI.46e). The separation distance between the two maxima depends on the specimen thickness as shown on figure XI.46f

From these calculated rocking curves, we conclude that the splitting of the diffracted maximum into two maxima with equal intensity will occur for any values of t, provided that the z values are not too far from zero. This splitting can hence be considered as typical and used to identify APBs with confidence.

XI.3.3.2.1 - Effect of an APB on CBED patterns

As explained in paragraph VII.5, the diffracted rocking curve is directly observed inside the diffracted disk of a CBED pattern along a diameter perpendicular to the excess line at $s = 0$. When the convergent incident beam is focused on a tilted APB, a symmetrical splitting of the superlattice line into two lines with equal intensity is expected provided that z is close to zero, i.e. the beam is located at the middle of the APB (Figure XI.47b). This behaviour is similar to that observed for stacking faults where the excess lines are asymmetrically split into a principal and a secondary line with different intensity.

The experimental patterns on figure XI.47 are in perfect agreement with these considerations.

XI.3.3.2.2 - Effect of an APB on LACBED patterns

The effect of an APB on LACBED patterns is more complex than that observed for CBED patterns. The difficulty lies in the fact that the LACBED method is a defocus method. Each point of the pattern corresponds to an incident ray and to a point of the specimen. In fact, for a tilted APB crossing the specimen, z varies along the Bragg line between the two limiting values $z = \pm t/2$ and s varies according to lines parallel to the excess line with $s = 0$. The rocking curves drawn between these two limiting values show that a superlattice Bragg line, undistorted in the perfect crystal, is split in the presence of an APB into two lines with equal intensity between $z = +t/2$ and $z = -t/2$, reaching the maximum separation when $z = 0$ (Figure XI.49d). This effect is, once more, similar to the one observed in the case of stacking faults where the excess lines are asymmetrically split into a principal and a secondary line with different intensity. The examples given on figure XI.49 are again in very good agreement with the theoretical calculations.

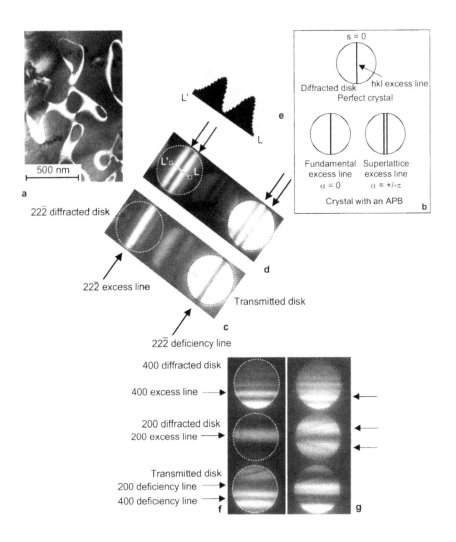

500 nm

a

222̄ diffracted disk

222̄ excess line

222̄ deficiency line

c

d

e

Transmitted disk

s = 0

Diffracted disk   hkl excess line

Perfect crystal

Fundamental   Superlattice
excess line   excess line
α = 0   α = +/-π

Crystal with an APB

b

400 diffracted disk

400 excess line

200 diffracted disk
200 excess line

Transmitted disk
200 deficiency line
400 deficiency line

f

g

Figure XI.47 - Effect of antiphase boundaries on CBED patterns.
a - Bright-field image of B2 ordered domains separated by APBs.
b - Description of the effect of an APB on fundamental and superlattice excess lines.

c - 222̄ dark-field CBED pattern taken inside a B2 ordered domain.

d, e - 222̄ dark-field CBED pattern taken on an APB separating two B2 ordered domains

and intensity profile through the splitting. The 222̄ excess and deficiency lines are split.
f - 200 dark-field CBED pattern taken inside a B2 ordered domain.
g - 200 dark-field CBED pattern taken on an APB. The 200 lines are split. Note that the 400
fundamental reflection is not affected.
Ni-Cu-Al shape-memory alloys. Courtesy of M.L. No and P. Rodriguez.

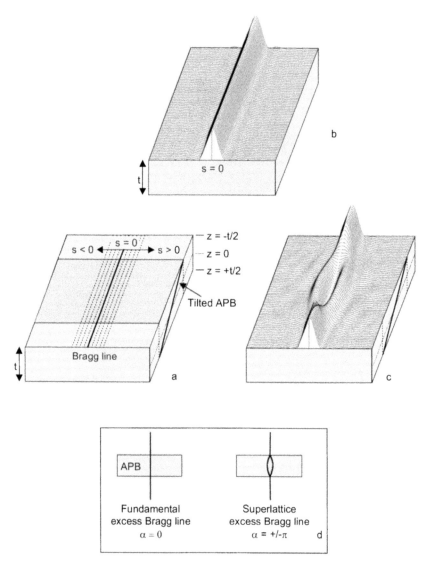

Figure XI.48 - Theoretical dark-field LACBED patterns under two beam conditions.
a - Principle of calculation of the effect of an APB on a superlattice excess Bragg line.
b - Rocking curves in the case of a perfect crystal.
c - Rocking curves drawn for z values between -t/2 and +t/2 in the case of a crystal containing a tilted APB. The rocking curves split into two maxima displaying the same intensity.
d - Schematic description of the effect of an APB on fundamental and superlattice excess Bragg lines.

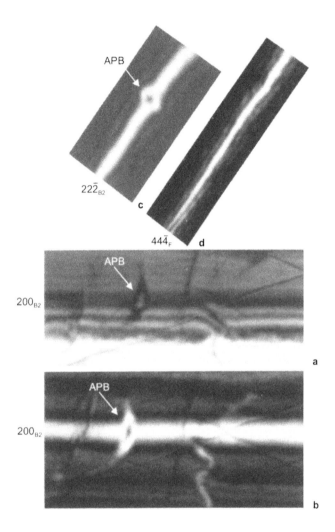

Figure XI.49 - Effect of antiphase boundaries on LACBED patterns.
a - Bright-field LACBED pattern showing the effect of an APB on the 200 deficiency Bragg line.
b - Corresponding 200 dark-field LACBED pattern. The effect of the APB is clearer.

c - $22\bar{2}$ dark-field LACBED pattern showing the effect of an APB.

d - $44\bar{4}$ dark-field LACBED pattern taken in the same area as c. The APB is not visible on this fundamental line.
Cu-Al-Ni shape-memory alloy. Courtesy of M.L. No and P. Rodriguez.

XI.3.4 - Characterization of grain boundaries

XI.3.4.1 - Crystallography of grain boundaries

A grain boundary is an interface separating two grains I and II, belonging either to the same phase (homophase boundary) or to different phases (heterophase boundary) (Figure XI.50a).

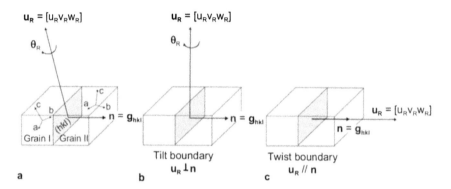

Figure XI.50 - Description of a homophase grain boundary.
a - Schematic description of a grain boundary.
b - Tilt boundary. The normal to the grain boundary **n** is perpendicular to the rotation axis u_R.
c - Twist boundary. The normal to the grain boundary **n** is parallel to the rotation axis u_R.

Such boundaries are characterized by:
- the orientation of the interface defined by (hkl) indices or by a vector **n** perpendicular to the boundary. This vector is a reciprocal lattice vector with:

n = g_{hkl} = ha* + kb* + lc*.
- the misorientation between the grains I and II. This is characterized by a rotation axis u_R and a rotation angle θ_R. The rotation axis u_R is a lattice direction $[u_R v_R w_R]$ = u_R**a** + v_R**b** + w_R**c**.

Depending on the relative orientation of the two vectors u_R and **n**, we can distinguish:
- twist boundaries when u_R is parallel to **n**,
- tilt boundaries when u_R is perpendicular to **n**,
- mixed boundaries when the vector u_R has an unspecified orientation with respect to **n**.

Depending on the value of the rotation angle θ_R, we can distinguish:

- large-angle boundaries when the rotation angle is larger than a few degrees,
- subgrain boundaries when the rotation angle θ_R is lower than 1 or 2°. In this case, the boundary can be described in terms of arrays of intrinsic dislocations and can be regarded as an internal crystal defect.

The misorientation can also be expressed in terms of a rotation matrix R describing the transformation of any direction $[u_I v_I w_I]$ from grain I into the corresponding direction $[u_{II} v_{II} w_{II}]$ in grain II.

$$R = \begin{vmatrix} r_{11} & r_{12} & r_{13} \\ r_{21} & r_{22} & r_{23} \\ r_{31} & r_{32} & r_{33} \end{vmatrix}$$

XI.3.4.1.1 - Particular cases

- Subgrain boundaries

Consider the subgrain tilt boundary shown on figure XI.51. This can be described in terms of an array of edge dislocations with dislocation lines parallel to the rotation axis (perpendicular to the page). These lines are separated by a distance D and they create a misorientation $\theta = b/D$. This angle becomes significant when the number of dislocations is large, that is to say, when the dislocations are separated by a very small distance. However, when the misorientation exceeds a few degrees, the distance D becomes smaller than the dislocation core and this type of description is no longer realistic. The dislocations responsible for the misorientation are called **primary intrinsic dislocations**. **Secondary intrinsic dislocations** are also observed and compensate for the misorientation deviations created by the primary dislocations. The boundary can also contain **extrinsic dislocations coming** from grains I and II. In the same way, a twist boundary can be regarded as an array of screw dislocations.

- Coincidence grain boundaries

In some special circumstances, the value of the misorientation of the two grains is such that some nodes of their respective lattices coincide to form a **coincidence site lattice (CSL)**. These coincidence boundaries are characterized by an index Σn where:
n = volume of the coincidence site lattice / volume of the initial lattice.

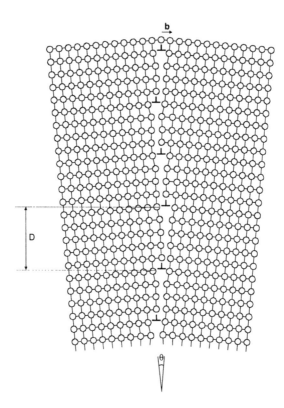

Figure XI.51 - Description of a tilt subgrain boundary.
This boundary is described in terms of an array of edge dislocation with Burgers vectors **b**. The dislocation lines are parallel to the rotation axis (perpendicular to the page) and are separated by a distance D. The misorientation θ is related to **b** and D by θ = b/D.

The example given on figure XI.52a corresponds to a Σ9 coincidence boundary present in a fcc structure. The two lattices are rotated by -38.94° around a common direction [011].

- Twin boundaries

Consider the special stacking sequence ABCABCBACBA... shown on figure XI.52b. The compact atomic plane C marked with an arrow separates, in a symmetrical manner, the lower and upper parts of the crystal. This plane is called a **twin plane** and the two corresponding lattices are rotated by 70.529° around the common [011] direction.

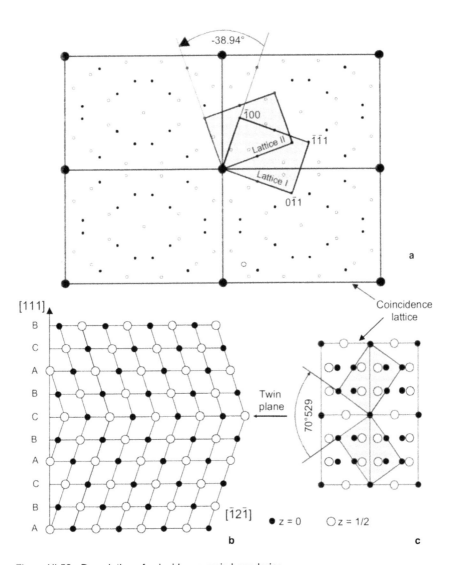

Figure XI.52 - Description of coincidence grain boundaries.

a - Description of a Σ9 coincidence grain boundary present in fcc structures. The two lattices are rotated by an angle of -38.94° around a common [011] direction.

b - Description of a twin boundary in fcc structures. The lower and upper parts of the crystal are symmetrical with respect to the twin plane.

c - Description of the twin boundary in term of a Σ3 coincidence grain boundary. The two lattices are rotated by an angle of 70.53° around a common [011] direction.

We note that this twin boundary is also a $\Sigma 3$ coincidence lattice (Figure XI.52c). This type of twin boundary is a **lattice merohedric twin** because the two grains have a common coincidence lattice.

Two other types of twin boundaries are also observed in crystals. These are:

- **merohedric twins**. This type of twinning occurs in crystals having a lattice more symmetrical than the crystal itself. In this case, the lattice is not disturbed by the twin boundary. This type of twin will be described in detail in paragraph XI.3.5.1 dealing with Dauphiné twins in quartz.

- **pseudo-merohedric twins**. These twins are observed in crystals having an orthorhombic lattice very close to a tetragonal lattice (pseudo-tetragonal crystal) or a quadratic lattice very close to a cubic lattice (pseudo-cubic crystal). Orthorhombic $YBa_2Cu_3O_7$ ceramics described in the paragraph XI.3.5.2 belong to this category.

XI.3.4.2 - Characterization of the grain boundary misorientation from LACBED patterns

In chapter X on pattern indexing, we have already indicated that any point of a LACBED pattern comes from an incident ray having a well-defined direction within the convergent incident beam. If we consider a point located on the grain boundary (for example the point A' on figure XI.53) we can identify:

- a direction $[u_I v_I w_I]$ for grain I,
- a direction $[u_{II} v_{II} w_{II}]$ for grain II

This means that a pair of parallel directions $[u_I v_I w_I] // [u_{II} v_{II} w_{II}]$ is obtained from the LACBED pattern. The choice of a point located on the grain boundary avoids any possible errors, which could otherwise be caused by a deformation or by a local relaxation of the specimen.

A second experiment is carried out with another orientation of the grain boundary with respect to the electron beam, in order to identify a second pair of parallel directions $[u'_I v'_I w'_I] // [u'_{II} v'_{II} w'_{II}]$.

Two pairs of parallel vectors $A_I // A_{II}$ and $A'_I // A'_{II}$ are obtained after normalisation of these experimental results (Figure XI.54a). The two cross-products $A_I \times A'_I$ and $A_{II} \times A'_{II}$ give C_I and C_{II} respectively (Figure XI.54b). The cross-products $C_I \times A_I$ and $C_{II} \times A_{II}$ give B_I and B_{II} (Figure XI.54c). Then, we can establish the two matrices $G_I = (A_I \ B_I \ C_I)$ and $G_{II} = (A_{II} \ B_{II} \ C_{II})$, which are related to the rotation matrix R by the equation:

$$G_{II} = R G_I \qquad\qquad (XI.4)$$

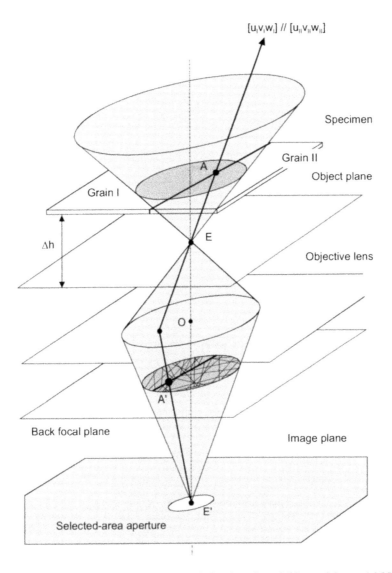

Figure XI.53 - Identification of a pair of parallel directions [u₁v₁w₁] // [u₁₁v₁₁w₁₁] from a LACBED pattern of a specimen containing a grain boundary.

Any point (for example the point A') of the LACBED pattern located on the grain boundary comes from an incident ray directed along the directions [u₁v₁w₁] and [u₁₁v₁₁w₁₁]. This pair of parallel directions can be identified from simulated diffraction patterns.

Thus, R can be identified from the two matrices G_I and G_{II} by the equation:

$$R = G_{II}G_I^{-1} \qquad (XI.5)$$

Finally, the rotation angle θ_R and the rotation axis u_R are obtained from the expressions:

$$\theta_R = \cos^{-1} 1/2(r_{11} + r_{22} + r_{33} - 1) \qquad (XI.6)$$

and

$$u = r_{32} - r_{23} / \sin\theta_R \qquad (XI.7)$$
$$v = r_{13} - r_{31} / \sin\theta_R \qquad (XI.8)$$
$$w = r_{21} - r_{12} / \sin\theta_R \qquad (XI.9)$$

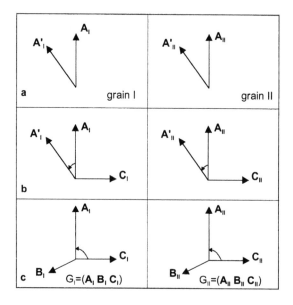

Figure XI.54 - Identification of the rotation matrix R from two pairs of parallel vectors A_I // A_{II} and A'_I // A'_{II}.
a - Disposition of the two pairs of parallel vectors A_I // A_{II} and A'_I // A'_{II} in grains I and II.
b - Identification of the vectors C_I and C_{II} from the cross-products A_I x A'_I and A_{II} x A'_{II}.
c - Identification of the vectors B_I and B_{II} from the cross-products C_I x A_I and C_{II} x A_{II}. The matrices G_I and G_{II} are established from the vectors A_I, B_I, C_I and A_{II}, B_{II}, C_{II}.

XI.3.4.2.1 - Application to the measurement of the misorientation of a $\Sigma 3$ twin boundary

The results of the analysis of a $\Sigma 3$ twin boundary present in a silicon specimen are described here to illustrate this method (Figure XI.55a) [XI.16].

The specimen is tilted so that the two following conditions are satisfied:

- the twin boundary is parallel or almost parallel to the incident beam axis. In this way, the transition between the two parts of the LACBED pattern located on either side of the grain boundary is narrow.

- the specimen orientation is chosen so that the LACBED pattern contains numerous sharp Bragg lines.

Figure XI.55b shows an experimental pattern obtained under these experimental conditions. We note that several deficiency lines (marked with an arrow) are continuous throughout the twin boundary. These lines correspond to common lattice planes belonging to the **DSC** (**D**isplacement **S**hift **C**omplete) lattice, which will be described in paragraph XI.3.6. The following pair of parallel directions is obtained from the point O chosen on the grain boundary using the theoretical diffraction pattern of figure XI.55c:

$[1142\ 185\ 1432]_I\ //\ [1212\ 1362\ 255]_{II}$

Figure XI.56a gives a second LACBED pattern from the same specimen area, obtained for a second crystal orientation. The same procedure gives a second pair of parallel directions:

$[\overline{254}\ 1461 1091]'_I\ //\ [\overline{332}\ 1169\ 1383]'_{II}$.

Note
The LACBED patterns corresponding to these two specimen orientations must be indexed in a coherent way.

The mathematical treatment described above leads to a rotation angle $\theta_R = 70.51° \pm 0.05°$ around the rotation axis $[\overline{4}\ 10^{-7}\ 707\ 707]$. These results are very close to the theoretical values for a $\Sigma3$ twin, namely a rotation of $70.529°$ around a $[011]$ axis.

The main point to consider is the accuracy in the measurement of the misorientation. This accuracy depends both on the quality of the LACBED patterns and on the sharpness of the lines involved in the analysis. The best accuracy is obtained with very sharp lines, i. e. quasi-kinematical lines having a large extinction distance. The sharpness of the lines can be further improved by energy filtering or by cooling the specimen. In the most favourable cases, the accuracy can reach $0.01°$.

The other identification techniques are less accurate than this. The best current methods, such as EBSD (Electron Back-Scattered Diffraction) or ECP (Electron Channelling Pattern) also use line patterns, but they contain broader lines than the sharp Bragg lines on LACBED patterns. The accuracy obtained with the best conventional methods does not exceed $0.1°$.

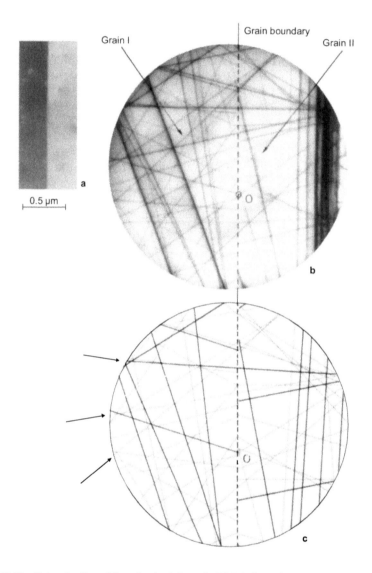

Figure XI.55 - Determination of the misorientation of a Σ3 twin boundary.
a - Micrograph of the Σ3 twin boundary.
b - LACBED pattern obtained for a first specimen orientation.
c - Corresponding simulated diffraction pattern. The lattice directions [1142 185 1432]I and [1212 1362 255]II are identified for the point O. The continuous lines marked with an arrow correspond to common lattice planes belonging to the DSC lattice (see chapter XI.3.6).
Pattern simulation carried out with the "Electron Diffraction" software [VI.3].
Silicon specimen. Courtesy of J. L. Maurice

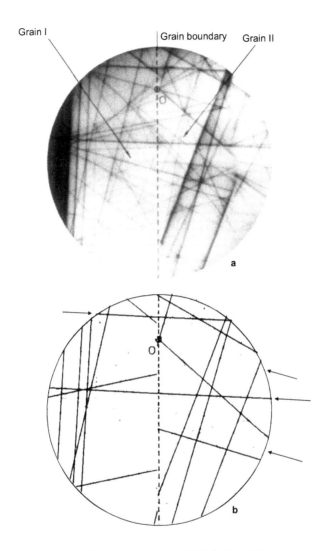

Figure XI.56 - Determination of the misorientation a Σ3 twin boundary.
a - LACBED pattern obtained for a second specimen orientation.

b - Corresponding theoretical diffraction pattern. The directions $[\overline{254}\ 1461\ 1091]'_I$ and

$[\overline{332}\ 1169\ 1383]'_{II}$ are identified for the point O.

The continuous lines marked by an arrow correspond to common lattice planes belonging to the DSC lattice (see chapter XI.2.4.3.3).

Simulation carried out with the "Electron Diffraction" software [VI.3].

Silicon specimen. Courtesy of J. L. Maurice.

- *Characterization of the grain boundary plane*

The identification of two directions $[u_1v_1w_1]$ and $[u_2v_2w_2]$ contained in the boundary allows the grain boundary plane to be characterized. We have indicated, in the paragraph X.5.2.1, that the cross-product of these two directions gives a reciprocal lattice vector $\mathbf{n} = \mathbf{g}_{hkl} = h\mathbf{a}^* + k\mathbf{b}^* + l\mathbf{c}^*$ perpendicular to the (hkl) boundary plane.

In the present case, the [011], [112] and [101] zone axes are observed when the twin plane is vertical. The cross-product of two of them, for example [011] x [112], gives a $(1\bar{1}1)$ twin boundary.

We have used this method to analyse $\Sigma 9$ and $\Sigma 25$ coincidence boundaries in silicon [XI.17] as well as twin boundaries in steels. Large angle boundaries have also been studied in α alumina [XI.18], steel and σ phase specimens.

XI.3.4.2.2 - Subgrain boundaries

The LACBED method remains valid for the study of very small misorientations, such as those observed in subgrain boundaries. We have applied it to a subgrain boundary present in the σ phase (Figure XI.57a). This subgrain boundary consists of an array of dislocations separated by a distance D = 0.1 ± 0.02 µm. The small misorientation associated with subgrain boundary is clearly visible on the $[\bar{1}\bar{1}5]$ zone-axis pattern (Figure XI.57b) where the central area of the LACBED pattern is sheared by the subgrain. Using the procedures described in the previous paragraph we obtain a rotation angle $\theta_R = 0.22° ± 0.05°$ around an axis very close to [010]. The LACBED analysis of dislocations shows that they are edge dislocations with a Burgers vector $\mathbf{b} = [001]$ and a line $\mathbf{u} = [010]$. The subgrain boundary is thus a tilt subgrain boundary [XI.19].

From the theoretical model given on figure XI.51 and the relation $\theta = b/D$, we obtain a rotation angle of 0.26° ± 0.05° around a direction [010]; this is in agreement with the experimental results.

Note
The Bragg lines used to identify the Burgers vector are extremely distorted owing to a strong deformation of the specimen area. Despite these distortions, Bragg lines can be indexed from the corresponding Kossel pattern by following the method described in paragraph VI.10.

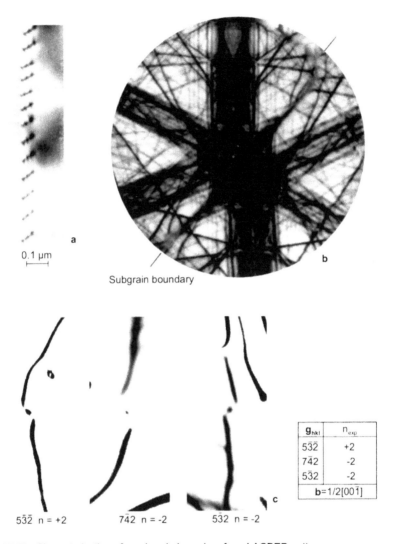

a

0.1 µm

Subgrain boundary

b

g_{hkl}	n_{exp}
$5\bar{3}\bar{2}$	+2
$7\bar{4}2$	-2
$5\bar{3}2$	-2
$b=1/2[00\bar{1}]$	

c

$5\bar{3}\bar{2}$ n = +2 $7\bar{4}2$ n = -2 $5\bar{3}2$ n = -2

Figure XI.57 - Characterization of a subgrain boundary from LACBED patterns.
a - Micrograph showing an array of edge dislocations separated by approximately 0.1µm.

b - Effect of the subgrain boundary on the central area of a $[\bar{1}15]$ zone axis LACBED pattern.

c - Effect of a subgrain boundary dislocation on the three Bragg lines used to identify the Burgers vector $b = [00\bar{1}]$. The Bragg lines are very distorted owing to a strong deformation of the specimen. The lines are indexed with the aid of the Kossel pattern and the procedure given in the paragraph VI.10.

Steel specimen. Courtesy of A. Redjaïmia.

XI.3.4.2.2 - Near-coincidence grain boundaries

Near coincidence grain boundaries contain arrays of dislocations connected with an additional misorientation from the perfect coincidence. This additional misorientation can be measured by the LACBED method described above but the accuracy is poor since the patterns are usually strongly affected by the dislocations present at the boundary and it is difficult to index accurately the lattice directions required for the analysis.

Another method consists in obtaining two LACBED patterns according to the experimental conditions given on figure XI.58 in the case of a near $\Sigma 11$ {311} coincidence grain boundary present in a nickel specimen (for a perfect $\Sigma 11$ grain boundary, the two grains are rotated by an angle of 50.47° around the common direction $S_0 = [101]$).

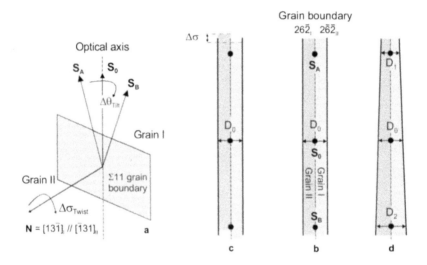

Figure XI.58 - Characterization of a near coincidence grain boundary from LACBED patterns.
a - Description of the experimental conditions used to characterize a near $\Sigma 11$ coincidence grain boundary present in a nickel specimen. The grain boundary is vertical in the microscope. A first LACBED pattern is obtained for the specimen orientation $S_A = [\bar{6}13]_I // [613]_{II}$ followed by a second pattern for the orientation $S_B = [316]_I // [3\bar{1}6]_{II}$.

b - Schematic disposition of the $26\bar{2}_I / \overline{26}2_{II}$ Bragg lines for the two specimen orientations S_A and S_B in the case of a perfect coincidence grain boundary. The two Bragg line are parallel and separated by a distance D_0.

c - Schematic disposition in the case of a coincidence grain boundary containing an additional twist component. The Bragg line remains parallel but the left part of the pattern is sheared by $\Delta\sigma$ with respect to the right part. The twist component $\Delta\theta_{twist}$ is obtained from $\Delta\sigma$.

d - Schematic disposition in the case of an additional tilt component. The two Bragg line are not parallel and the tilt component $\Delta\theta_{tilt}$ is obtained from $\Delta D = D_2 - D_1$.

Grain boundary

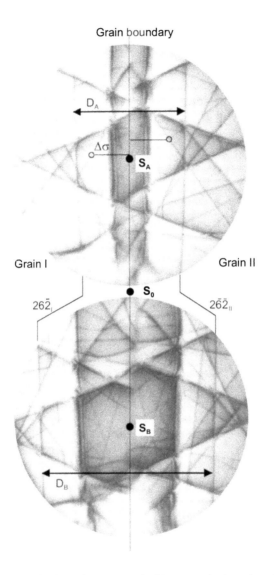

Grain I

Grain II

262_I

262_{II}

Figure XI.59 - Characterization of a near coincidence grain boundary from LACBED patterns. LACBED patterns obtained for the two symmetrical crystal orientations $S_A = [6\bar{1}3]_I \,//\, [613]_{II}$ and $S_B = [3\bar{1}6]_I \,//\, [316]_{II}$. The distances D_A and D_B, which separate the two Bragg lines $26\bar{2}/2\bar{6}\bar{2}$ on the patterns S_A and S_B, are different. The left and right parts of the patterns are sheared by a distance $\Delta\sigma$. A twist component $\Delta\theta_{twist} = 1.31 \,!\, 0.02°$ and a tilt component $\Delta\theta_{tilt} = 0.16 \,!\, 0.02°$ are inferred from these two patterns.
Nickel bi-crystal. Courtesy of E. Priester and S. Poulat.

On the two LACBED patterns thus obtained, the distances D_A and D_B separating the two symmetrical Bragg lines $26\overline{2}_I/\overline{26}\overline{2}_{II}$ on either side of the boundary are different. In agreement with the theoretical models on figure XI.58d, this means that this near coincidence grain boundary contains an additional tilt component $\Delta\theta_{tilt} = 0.16°$! 0.02°. In addition, the two patterns also display a shear $\Delta\sigma$ of the left part with respect to the right one. This is typical of a twist component $\Delta\theta_{twist} = 1.31°$! 0.02°.

These accurate values can be correlated with the dislocation arrays [XI.19].

XI.3.5 - Twins

Two other interesting applications of the LACBED technique are presented in this paragraph. They concern Dauphiné twins in quartz and twin boundaries in ceramic superconductors.

XI.3.5.1. - Dauphiné twins

Dauphiné twins are merohedric twins observed in α quartz [XI.10]. They are related to the fact that α quartz has a trigonal structure (characterized by a 3-fold rotation axis) and belongs to the space group $P3_221$ for the left-hand variety and $P3_121$ for the right-hand variety. Both varieties have the same hexagonal lattice, characterized by a 6-fold rotation axis (Figures XI.60a and c).

Dauphiné twins separate two domains rotated by a 180° rotation around the [0001] axis. Figures XI.60b and d show that this rotation changes the atomic positions in the unit cell but does not modify the lattice. The (hkil) lattice planes are transformed into $(\overline{h}\overline{k}i\overline{l})$ planes. A 60° or 300° rotation around the [0001] direction would produce the same effects. From this model, we anticipate that all Bragg lines on the LACBED pattern will be continuous across the twin boundary since the position of the lines depends only on the lattice. On the other hand, their intensity should change at the twin boundary since the intensity of the lines depends on the extinction distance ξ_g. The latter is related, via the structure factor F_{hkl}, to the nature and to the position of the atoms in the unit cell.

These properties are actually observed on the experimental pattern on figure XI.61a. It is in perfect agreement with the theoretical pattern of figure XI.61b where the hkil indices as well as the extinction distances ξ_g of the main Bragg lines are given. Note that the $(3\overline{3}61)$ and

(33$\bar{6}$1) lines display the same intensity because they have the same extinction distance. Some lines seem to stop at the boundary. In reality, these lines are continuous but their intensity on one side of the twin is too low to be observed.

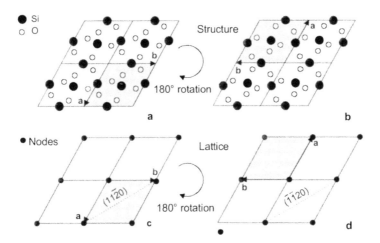

Figure XI.60 - Characterization of Dauphiné twins in α quartz
a - Projection of the α quartz structure along the [0001] direction.
b - Structure obtained after a180° rotation around the [0001] direction. The atomic positions are changed.
c - Projection of the lattice along the [0001] direction.
d - Lattice obtained after a 180° rotation around the [0001] direction. The positions of the lattice nodes are not modified. The (hkil) lattice planes are transformed into the ($\bar{h}\bar{k}$il) lattice planes. The example of the (11$\bar{2}$0) planes transformed into ($\bar{1}\bar{1}$20) planes is indicated on the figure.
Note that a rotation of 60° or 300° around the [0001] direction would produce the same effects.
Simulation carried out with the "Electron Diffraction" software [VI.3].

XI.3.5.2 - Characterization of twin boundaries in ceramic superconductors

Figure XI.63a shows an YBaCuO specimen containing a large number of microtwins. These pseudo-merohedric twins occur very easily because the YBaCuO orthorhombic structure (with lattice parameters a = 0.382 nm, b = 0.388 nm and c = 1.168 nm) is almost tetragonal. Figure XI.62a shows that a (110) twin boundary results from a rotation of the two lattices by an angle φ around the common [001] direction.

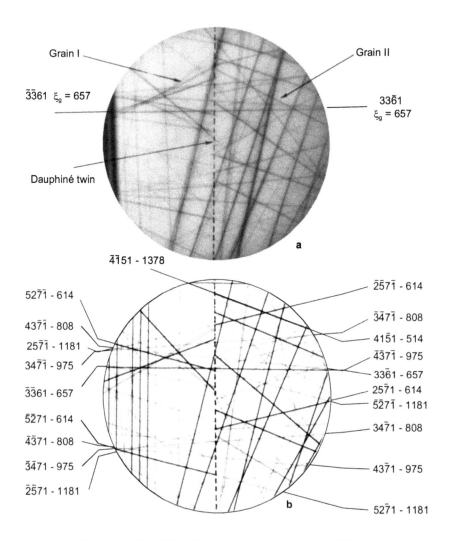

Figure XI.61 - Characterization of Dauphiné twins in α quartz from LACBED patterns.
a - Experimental LACBED pattern. All the lines are continuous but their intensity changes at the twin boundary. Note that the $33\bar{6}1$ and $\overline{33}61$ lines are not affected. Some lines seem to stop at the interface. Actually, these lines are continuous but their intensity on one side of the twin is too low to be observed.
b - Simulated LACBED pattern. The hkil indices as well as the extinction distances ξ_g (given in nm and for 300 kV) of the main Bragg lines are indicated. A large extinction distance corresponds to a weak intensity.
Simulation carried out with the software "Electron Diffraction" [VI.3].
Quartz LACBED pattern. Courtesy of P. Cordier.

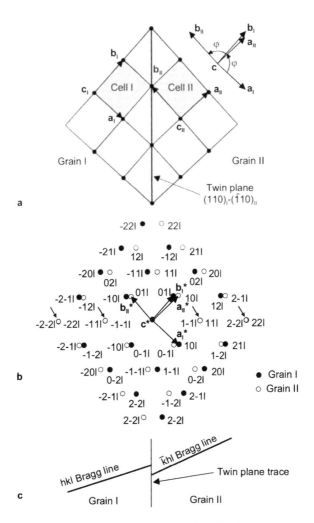

Figure XI.62 - Pseudo-merohedric twins present in an $YBa_2Cu_3O_7$ specimen.

a - Schematic description of the twinning. The two lattices from grains I and II are rotated by an angle ϕ around the common [001] direction.

b - Projection of the two reciprocal lattices along c*. The hkl nodes from grain I are very close to the $\bar{k}hl$ nodes from grain II. The $h\bar{h}l$ and hhl nodes (marked with an arrow) are perfectly superimposed.

c - Schematic description of the arrangement of a pair of $hkl/\bar{k}hl$ Bragg lines at the intersection with a twin boundary.

For the sake of clarity, the slight difference between the two lattice parameters a and b is deliberately exaggerated on these drawings.

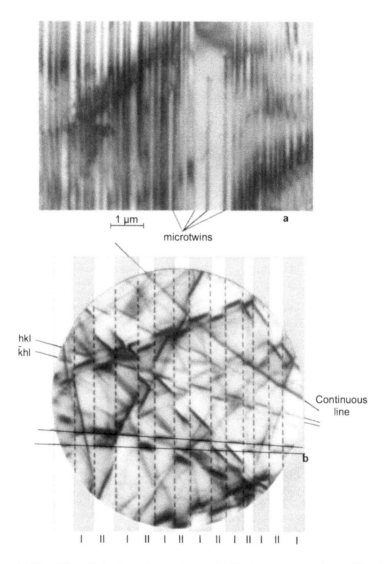

Figure XI.63 - Characterization of pseudo-merohedric twins present in an $YBa_2Cu_3O_7$ specimen.

a - Bright-field micrograph showing numerous microtwins.

b - Corresponding LACBED pattern. The pairs of of $hkl/\bar{k}hl$ Bragg lines are discontinuous at each intersection with the twin domains (denoted I and II). Note the presence of a continuous line corresponding to a common set of lattice planes.

$YBa_2Cu_3O_7$ patterns. Courtesy of J. Ayache.

This angle ϕ is related to the b/a ratio by $\phi = 2\tan^{-1}(b/a)$. In the present case $\phi = 90.892°$, which is very close to $\pi/2$. Since the two reciprocal lattices undergo the same misorientation as their direct lattices (Figure XI.63b), each hkl node from grain I is located very close to the $\bar{k}$hl node from grain II. The nodes hhl and h$\bar{h}$l (indicated by arrows on figure XI.62) are perfectly superimposed. They correspond to sets of lattice planes not affected by the twin boundary. As a result, the pairs of hkl/$\bar{k}$hl Bragg lines display a discontinuity and a small misorientation when they intersect a twin boundary (Figure XI.62c). A detailed description of these effects has been given by Tanaka *et al.* [IX.3].

The pairs of Bragg lines are discontinuous at each intersection with the microtwin boundaries (figure XI.63b). Midgley *et al.* characterized the b/a ratio accurately by measuring this effect [XI.20].

In order to show the effects of these twins more clearly, the very slight difference between the lattice parameters a and b is deliberately exaggerated on figure XI.62. In practice, the pairs of hkl/$\bar{k}$hl nodes are very close to one another and their separation is very difficult to observe by conventional diffraction techniques. This difficulty can be overcome with the LACBED technique owing to its high sensitivity to very small lattice plane misorientations.

XI.3.6 - Characterization of grain boundary dislocations

The $\Sigma 3$ twin boundary, studied in paragraph XI.2.4.3.1, contains some secondary intrinsic dislocations, which compensate for the departure from perfect coincidence. One of these dislocations is clearly visible on figure XI.64a. The Burgers vector of these dislocations can also be identified from LACBED patterns provided that Bragg lines common to both grains are chosen. These lines correspond to lattice planes belonging to the DSC (Displacement Shift Complete) lattice. The DSC lattice is the smallest lattice that takes into account the lattice nodes from both lattices. We can expect that grain boundary dislocations will have Burgers vectors in agreement with the vectors of this DSC lattice.

Figure XI.64b shows the DSC lattice corresponding to a $\Sigma 3$ twin boundary. In addition to the usual dislocations with $\mathbf{b} = 1/2<110>$, dislocations with $\mathbf{b} = 1/6<112>$ and $\mathbf{b} = 1/3<111>$ are also expected. In this case, integer values of the $g_{hkl}.\mathbf{b}$ dot product should be obtained for the common Bragg lines. The experimental patterns obtained in this way show that this is indeed the case.

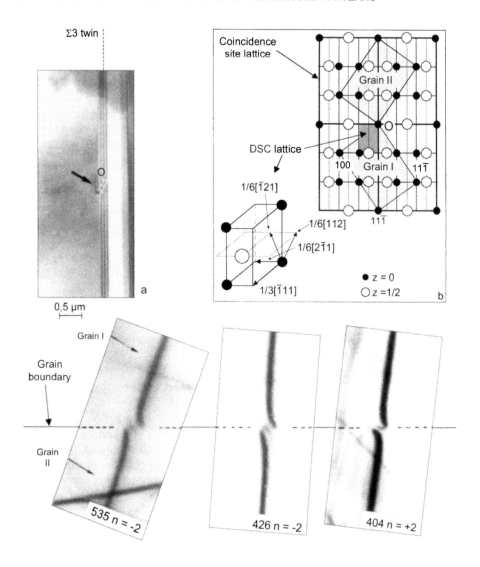

Figure XI.64 - Characterization of a dislocation in a Σ3 coincidence grain boundary.
a - Micrograph of the twin boundary containing a dislocation.
b - Description of the DSC lattice for a Σ3 coincidence boundary giving the possible Burgers vectors for boundary dislocations.

c - Effects of the dislocation on three common Bragg lines. The Burgers vector **b** = 1/6 [2$\bar{1}\bar{1}$] is in agreement with the DSC lattice. All the hkl and uvw indices are given with respect to the grain I, which is used as reference.
Silicon specimen. Courtesy of J. L. Maurice.

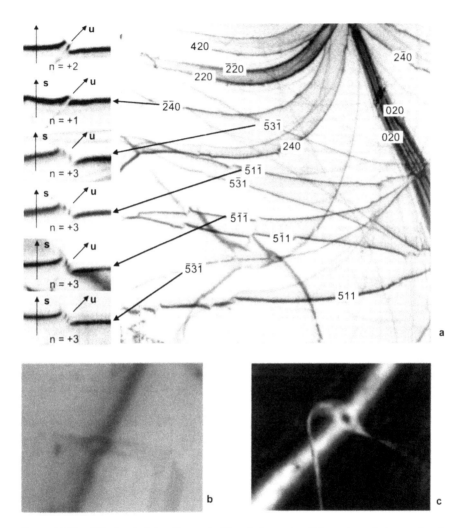

Figure XI.65 - Effects of dislocations, stacking faults and antiphase boundaries on bend-contour patterns.

a - Identification of the Burgers vector of a dislocation in a stainless steel specimen from bend contour patterns. The splittings at the intersection of the dislocation line with the bend contours are similar to those observed on LACBED patterns. They lead to $\mathbf{b} = 1/2\,[\bar{1}0\bar{1}]$.

b - Effect of a stacking fault on a bend contour. The bend contour is split into a principal and a secondary contour.

c - Effect of an antiphase boundary on a bend contour. The bend contour is split into two contours having the same intensity.

An example of such an analysis is given on figure XI.64c. It leads to a Burgers vector $\mathbf{b} = 1/6\,[2\bar{1}1]$ in perfect agreement with the DSC lattice described above. Other examples are given in reference [XI.21].

XI.4 - Characterization of crystal defects from bend contour patterns

In chapter VI we mentioned the very strong analogy between bend-contour and LACBED patterns. Therefore, we can expect to observe similar effects at the intersection of defects with bend contours and Bragg lines. Figure XI.65 proves that it is indeed the case.

Figure XI.65a gives a complete analysis of a dislocation present in a ferrite grain of a steel specimen [XI.22]. The thin foil was very distorted near the edge where many bend contours were observed. Many nice splittings were obtained by slightly tilting the specimen so that the dislocation line intersects different bend contours. They lead to the Burgers vector $\mathbf{b} = 1/2\,[\bar{1}0\bar{1}]$.

This type of experiments can be useful in the case of a specimen containing a high dislocation density. Isolated dislocations suitable for the analysis may be found in the thinnest specimen areas near the edge where bend contours are frequently present. Of course, it is important to ensure that these dislocations are really typical of the bulk specimen. This method is very easy to perform since it is realized in image mode with a specimen located at the eucentric height. The specimen is tilted so that the dislocation line moves from one bend contour to another.

In the same way, stacking faults and antiphase boundaries also produce typical effects on bend contours [XI.24]. The two examples shown on figures XI.65b and c prove that the effects are similar to those observed on LACBED patterns.

CONCLUSION

LACBED is, without a doubt, the richest of the electron diffraction techniques for LACBED patterns can contain several hundred Bragg lines. These lines can be grouped into pairs of parallel lines: the distance and the angle between them are directly related to the lattice spacings and to the angles between the corresponding diffracting lattice planes. Their intensities are dependent, in a more or less direct way, on the structure factor.

LACBED patterns display an excellent quality because of the efficient filtering of the scattered electrons by the selected-area aperture. Owing to the defocused mode used in this technique, the patterns contain simultaneously several Bragg lines and the shadow image of the illuminated area of the specimen. The presence, on the same pattern, of information on both the reciprocal and the direct spaces constitutes the wealth and the originality of the LACBED technique. LACBED patterns are thus image-diffraction mappings and these mappings are very sensitive to local modifications of the diffraction conditions such as those due to crystal defects. New fields of applications become accessible and were first explored by Cherns and Preston in 1986, who reported the typical effects of dislocations on Bragg lines [i.7]. As these effects are directly related to the $g_{hkl}.b$ dot products, LACBED is a more powerful Burgers vector identification method than other methods based on invisibility criteria. Here, we observe obvious and clear effects because the two-beam conditions are locally satisfied and the image is improved by the angular filtering of the selected-area aperture. In some cases, a single pattern is sufficient for reliable identification of the Burgers vectors. Usually, two or three patterns are required and it is necessary neither to tilt the specimen (because the LACBED patterns always contain enough useful lines) nor to make any assumptions on the nature of the Burgers vectors. The method is valid with any types of dislocations: perfect dislocations, partial dislocations, grain boundary dislocations, dislocations in quasicrystals... LACBED also remains valid for the analysis of anisotropic materials, where the traditional methods are difficult to use.

LACBED has also been applied successfully to planar defects such as stacking faults, antiphase boundaries and grain boundaries. Grain boundary misorientations can be measured with an accuracy of about 0.05°, which cannot be reached with other techniques. This performance is possible because the measurements are made from the sharp and weak Bragg lines present on the patterns.

A very interesting field of applications concerns the measurement of local strains. This is confirmed by the large number of publications on this subject.

LACBED has yet another interesting property. It is a "low-dose" technique since the defocus illumination mode spreads the electrons of the incident beam over a relatively large surface of the specimen. Applications to organic and quartz crystals that are very sensitive to electron irradiation exploit this property. There is no doubt that applications to other sensitive materials will follow.

Alternatives to LACBED are also valuable, especially the CBIM and the defocus CBED variants. In fact, these variants preserve the eucentric specimen position even in the defocused mode. This property is very convenient for the microscopist. Moreover, we can switch easily from a conventional image to a line pattern with the CBIM method, or from a focused CBED pattern to the defocused case. We just need to modify slightly the operating current of the C_2 condenser.

It is relatively easy to obtain LACBED patterns, as long as the microscope operating modes are well known and understood. Certainly, it is necessary to work with strongly convergent beams and with very small probe sizes and the effects of the spherical aberration become very significant. Nevertheless, in many cases, poor adjustment of the microscope affects neither the intrinsic quality of the patterns nor the information that they contain. This is a great advantage.

This review would be incomplete without some comment on the drawbacks. The major difficulty concerns the large size of the area illuminated. This problem can be solved, to a large extent, by using a very small selected-area aperture (1, 2 or 5 µm) and a small height Δh. In this way, the size of the illuminated area can be reduced to less than 0.1 µm.

A second difficulty is connected with pattern distortion. LACBED is very sensitive to specimen deformations that distort the patterns. Most of the patterns shown in this book display few distortions because they come from specimens prepared by the "tripod" technique [c.1]. This technique produces very flat specimens, the thickness of which increases very slowly. We recommend that it be used whenever the nature of the specimen permits it.

Many detailed drawings illustrate this text and three-dimensional representations have been used whenever possible. The <001> silicon zone-axis pattern is often chosen for the experimental examples. It has the major advantage of allowing easy comparison between the various techniques and of highlighting the effects of the various experimental parameters studied in this book.

In this monograph, I give a personal view that may occasionally be over-simplified. This is the result of many years of experimental observations, which have shown that most of the qualitative aspects of diffraction patterns can be interpreted on this basis.

What are the perspectives?
All the descriptions given in this book are concerned mainly with qualitative aspects. For example, we simply count the number of fringes at the intersection of the dislocation lines with the Bragg lines in order to identify the Burgers vector of dislocations. More detailed analyses would take into account the quantitative aspects of diffraction, i.e.measurements of the transmitted and diffracted intensities. Two conditions are then required:
- the inelastic scattering that disturbs the diffracted intensities and produces the background must be removed. The angular filtering of the selected-area aperture eliminates most of these inelastic electrons. We can further improve this filtering by means of an energy filter.
- the diffracted intensities must be measured accurately. This is now possible with CCD cameras or with devices such as "Imaging Plates".
This opens a new field of quantitative (large-angle) convergent-beam electron diffraction, which constitutes a major research field worldwide.

REFERENCES

i.1 A new electron microscope with continously variable magnification
J.B. LE POOLE
Philips Technical Review
1947, 9, 33

i.2 Elektroneninterferenzen im konvergenten Bundel
W. KOSSEL and G. MÖLLENSTEDT
Annalen der Physik
1939, 36, 113

i.3 The symmetry of electron diffraction zone axis patterns
B.F. BUXTON, J.A. EADES, J.W. STEEDS and G.M. RACKHAM
Philosophical Transactions of the Royal Society of London
1976, A281, 181-194

i.4 Extinction conditions in the dynamical theory of electron diffraction
J. GJØNNES and A.F. MOODIE
Acta Crystallographica
1965, 19, 65

i.5 The determination of foil thickness by scanning transmission electron microscopy
P.M. KELLY, A. JOSTSONS, R.G. BLAKE and J.G. NAPIER
Physica Status Solidi
1975, A31, 771-779

i.6 Large Angle CBED
M. TANAKA, R. SAITO, K. UENO and Y. HARADA
Journal of Electron Microscopy
1980, 29, 408-412

i.7 Convergent beam studies of crystal defects
D. CHERNS and A.R. PRESTON
Proceedings ICEM-11, Kyoto
1986, 721-722

i.8 Formation simultanée de deux diagrammes de diffraction électronique
J. BEAUVILLAIN
Journal de Microscopie
1970, 4, 455-464

REFERENCES

i.9 Line patterns in wide angle convergent beam electron diffraction
D.J. SMITH and J.M. COWLEY
Journal of Applied Crystallography
1971, 4, 482

III.1 Über die Genauigkeit der Übereinstimmung von ausgewälten und beugenden Bereich
bei der Feinbereichs-Elektronenbeugung im Le Pooleschen Strahlengang
W.D.RIECKE
Optik
1961, 18, 278-293

III.2 Microdiffraction as a tool for crystal structure identification and determination
J.P. MORNIROLI and J.W STEEDS
Ultramicroscopy
1992, 45, 219-239

VI.1 Effect of energy filtering in LACBED patterns
I.K. JORDAN, C.J. ROSSOUW and R. VINCENT
Ultramicroscopy
1991, 35, 237-243

VI.2 International Tables for Crystallography
Reidel, Dordrecht
1983, vol. 4

VI.3 "Electron Diffraction".
J.P. MORNIROLI, D. VANKIEKEN et L. WINTER
A sofware for the simulation of electron diffraction patterns
Laboratoire de Métallurgie Physique et Génie des Matériaux, UMR CNRS 8517, USTL

VI.4 Real space crystallography
J.W STEEDS, G.J TATLOCK and J. HAMPSON
Nature
1973, 435-439

VII.1 Electron Microscopy of Thin Crystals
P.B. HIRSCH, A. HOWIE, R.B. NICHOLSON, D.W. PASHLEY and M.J. WHELAN
1975, Butterworth, London

VII.2 Coherent overlapping LACBED patterns in 6H SiC
R. VINCENT, P.A. MIDGLEY, P. SPELLWARD and J.W. STEEDS
Ultramicroscopy
1993, 50, 365-376

VII.3 Observation of lattice fringes in CBED
M. TERAUCHI, K. TSUDA, O. KAMIMURA, M. TANAKA, T. KANEYAMA and T. HONDA
Ultramicroscopy
1994, 54, 268-275

VII.4 Electron Microdiffraction
J.C.H. SPENCE and J.M. ZUO
Plenum, New York & London
1992

VII.5 Quantitative electron microdiffraction
J.C.H. SPENCE
Journal of Electron Microscopy
1996, 45, 19-26

VII.6 Different types of HOLZ-line interactions and their use in structure factor determination
R.M.J. BOKEL, F.W. SCHAPINK and F.D. THICHELAAR
Ultramicroscopy
1996, 65, 1-12

VII.7 Crystallographic data from zone axis patterns
J.W. STEEDS, J.R. BAKER and R. VINCENT
Proceedings ICEM-10, Hamburg
1982, 617-624

VII.8 Microscopia Elettronica in Trasmissione e Techniche di Analisi di Superfici nella
Scienza dei Materiali. B, Microscopia Elettronica in Trasmissione
Convergent Beam Electron Diffraction
J.W STEEDS
Edizione ENEA, Roma
1986, 256-257

VII.9 Identification of forbidden reflections by LACBED
J.P. MORNIROLI, M. MANGOLD and O. RICHARD
Proceedings. ICEM 13, Paris
1994, 915-916

VIII.1 Techniques of CBED
R. VINCENT
Journal of Electron Microscopy Technique
1989, 13, 40-50

IX.1 Convergent beam electron diffraction and imaging of strained layer superlattices
D.M. MAYER, H.L. FRASER, C.J. HUMPHREYS, R.V. KNOELL, R.D. FIELD, J.B.
WOODHOUSE and J.C. BEAN
Electron Microscopy and Analysis, IOP Conference Series
1986, 78, 49-52

IX.2 Convergent-beam imaging - a transmission electron microscopy technique for
investigating small localized distorsions in crystals
C.J. HUMPHREYS, D.M. MAHER, H.L. FRAZER, D. EAGLESHAM
Philosophical Magazine
1988, 58, 787-798

IX3 - Application of the Convergent Beam Imaging (CBIM) technique to the analysis of
crystal defects
J.P. MORNIROLI, P. CORDIER, E. VAN CAPPELLEN, J.M. ZUO and J. SPENCE
Microscopy, Microanalysis, Microstructures
1997, 8, 187-202

IX.4 Identification of lattice defects by CBED
M. TANAKA, M. TERAUCHI and T. KANEYAMA

REFERENCES

Journal of Electron Microscopy
1991, 40, 211-220

IX.5 Determination of the displacement vector of a stacking fault in TiO_2 by CBED
S. YAMADA, M. TANAKA
Journal of Electron Microscopy
1995, 44, 212-218

IX.6 Simultaneous observation of bright field and dark field LACBED patterns
M. TERAUCHI and M. TANAKA
Journal of Electron Microscopy
1985, 34, 128-135

IX.7 Simultaneous observation of zone axis patterns and ± g dark field patterns in CBED
M. TERAUCHI and M. TANAKA
Journal of Electron Microscopy
1985, 34, 347-356

IX.8 Zone-axis patterns formed by a new double-rocking technique
J.A. EADES
Ultramicroscopy
1980, 5, 71-74

IX 9 Electron diffraction with rocking beam and rocking crystal (analogies: CTEM-STEM)
M. BRUNNER, H.J. KOHL and N. NIEDRIG
Optik
1981, 58, 37-55

IX.10 Direct parallel detection of energy-resolved large angle convergent-beam patterns
E. QUANDT, S. LA BARRE and H. NIEDRIG
Ultramicroscopy
1990, 33, 15-21

IX.11 Specimen contamination in the electron microscope when small probes are used
G.M. RACKHAM and J.A. EADES
Optik
1977, 47, 227-232

IX.12 Refinement of crystal structure using CBED : the low-temperature phase of $SrTiO_3$.
K. TSUDA and M. TANAKA
Acta Crystallographica
1995, A51, 7-19

IX.13 Quantitative electron microdiffraction
J.C.H. SPENCE
Journal of Electron Microscopy
1996, 45, 19-26

X.1 EMS - A software package for electron diffraction analysis and HREM simulation in materials science
P.A. Stadelmann
Ultramicroscopy
1987, 21, 131-146

X.2 Trace analyses from LACBED patterns
J.P. MORNIROLI and F. GAILLOT
Ultramicroscopy
2000, 3, 227-243

X.3 Determination of the orientation of a stacking fault by LACBED
X. WEI, X. DUAN and S. WAN
Ultramicroscopy
1996, 66, 49-57

XI.1 Analyses par LACBED de bulles d'hélium dans des alliages $Pd_{90}Pt_{10}$ tritiés
J.P. MORNIROLI, B. DECAMPS and D. JACOB
Actes du IVe Colloque de la SFμ, Toulouse
2000, 74

XI.2 - Microscopia Elettronica in Trasmissione e Techniche di Analisi di Superfici nella
Scienza dei Materiali. B, Microscopia Elettronica in Trasmissione
Convergent Beam Electron Diffraction
J.W STEEDS
Edizione ENEA, Roma
1986

XI.3 Convergent beam diffraction studies of interfaces, defects and multilayers
D. CHERNS and A.R. PRESTON
Journal of Electron Microscopy Technique
1989, 13, 111-122

XI.4 CBED
M. TANAKA and M. TERAUCHI
vol. II, JEOL, Tokyo
1988

XI.5 CBED
M. TANAKA, M. TERAUCHI and K. TSUDA
vol. III, JEOL, Tokyo
1994

XI.6 TEM investigation of dislocation microstructure of experimentally deformed silicate
garnet
P. CORDIER, P. RATERRON and Y. WANG
Physics of the Earth and Planetary Interiors
1996, 97, 121-131

XI 7 LACBED and CBIM analysis of dislocations in anisotropic materials
J.P. MORNIROLI, J. PONS and R. PORTIER
Proceedings EUREM-11, Dublin
1996

XI.8 Characterization of dislocations in anisotropic materials by LACBED
J.P. MORNIROLI and P. VERMAUT
Journal de Physique IV
1993, 3, 2165-2168

REFERENCES

XI.9 CBED from organic crystals
R. VINCENT
IOP Conference Series
1985, 78, 427-428

XI 10 Characterization of crystal defects in quartz by LACBED
P. CORDIER, J.P. MORNIROLI and D. CHERNS
Philosophical Magazine A
1995, 72, 1421-1430

XI.11 Characterization of the Burgers vectors of partial dislocations by LACBED
J.P. MORNIROLI and D. CHERNS
IOP Conference Series
1993, 138, 153-156

XI.12 Computed Electron Micrographs and Defect Identification
A.K. HEAD, P. HUMBLE, L.M. CLAREBROUGH, A.J. MORTON and C.T. FORWOOD
North-Holland, Amsterdam
1973

XI.13 Distortion of the zeroth-order Laue zone pattern caused by dislocations in a silicon crystal.
J. WEN, R. WANG and G. LU
Acta Crystallographica
1989, A45, 422-427

XI.14 CBED characterization of dislocations in GaS single crystals.
C. DE BLASI, D. MANNO, A. RIZZO
Ultramicroscopy
1990 33, 143-149

XI.15 CBED and LACBED characterization of antiphase boundaries
J.P. MORNIROLI, M.L. NO, P.P. RODRIGUEZ, J.SAN JUAN, E. JEZIERSKA, N. MICHEL, S. POULAT and L. PRIESTER
Submitted to Ultramicroscopy

XI.16 Accurate measurement of grain boundary misorientation by LACBED
J.P. MORNIROLI
Interface Science
1997, 4, 273-283

XI.17 LACBED studies of grain boundaries
J.P. MORNIROLI, F. STRZELCZYK, A. REDJAIMIA and D. CHERNS
Proceedings. ICEM-13, Paris
1994, 1, 901-902

XI.18 Characterization of grain boundary misorientation in α alumina by LACBED
J.P. MORNIROLI, A. LECLERE, W. SWIATNICKI and J.Y. LAVAL
Journal de Physique IV
1993, 3, 1455-1458

XI.19 Grain boundary dislocation network characterization by Large-Angle Convergent-Beam Electron Diffraction (LACBED) associated with the weak beam technique

S. POULAT, J.P. MORNIROLI and L. PRIESTER
Proceedings X Conference on Electron Microscopy of Solids, Warsaw-Serock, Poland
1999, 151-155

XI.20 {110} twinning in $YBa_2Cu_3O_{7-x}$
P.A. MIDGLEY, R. VINCENT, D. CHERNS
Philosophical Magazine A
1992, 66, 237-256

XI.21 Analysis of grain boundary dislocations by LACBED
J.P. MORNIROLI and D. CHERNS
Ultramicroscopy
1996, 62, 53-63

XI.22 Characterization of the Burgers vector of dislocations from bend contours
J.P. MORNIROLI and P. CORDIER
Proceedings EUREM-12, Brno
2000, 2, 507-508

c.1 A procedure for cross sectioning materials for TEM analysis without ion milling
J.P. BENEDICT, R. ANDERSON, S.J. KLEPEIS and M. CHAKER
Materials Research Society Symposium Proceedings.
1990, 199, 189

BIBLIOGRAPHY

General references (CBED, LACBED)

Practical Analytical Electron Microscopy in Materials Science
D.B. WILLIAMS
Philips Electronic Instruments, Electron Optics Publishing Group
1984

Convergent Beam Electron Diffraction of Alloy Phases
The Bristol Group (Compiled by J. Mansfield)
Adam Hilger, Bristol
1984

Electron Beam Analysis of Materials
M.H. LORETTO
Chapman and Hall, London
1984

Selected Area Electron Diffraction (SAED) and Convergent Beam Electron Diffraction
J.W. STEEDS and J.P. MORNIROLI
In Minerals and Reactions at the Atomic Scale. Transmission Electron Microscopy
P.BUSECK, Ed
Reviews in Mineralogy
1992, 27, 37-84

Convergent Beam Electron Diffraction
M. TANAKA and M. TERAUCHI
vol. I, JEOL, Tokyo, 1985
M. TANAKA and M. TERAUCHI and T. KANEYAMA
vol. II, JEOL, Tokyo, 1988
M. TANAKA, M. TERAUCHI and T. TSUDA
vol. III, JEOL, Tokyo, 1994
M. TANAKA, M. TERAUCHI, T. TSUDA and K. SAITOH
vol. IV, JEOL, Tokyo, 2002

Convergent Beam Electron Diffraction
J.W STEEDS
In Microscopia Elettronica in Trasmissione e Techniche di Analisi di Superfici nella Scienza
dei Materiali. B, Microscopia Elettronica in Trasmissione
Edizione ENEA, Roma
1986

BIBLIOGRAPHY

Electron Microdiffraction
J.C.H. SPENCE and J.M. ZUO
Plenum, New York & London
1992

Electron Diffraction Techniques
J.M. COWLEY Ed.
IUCr Monographs on Crystallography, Oxford University Press
1992

Electrons et Microscopes
P. HAWKES, Ed.
CNRS Editions, Belin, Paris
1995

Transmission Electron Microscopy
D.B. WILLIAMS and C.B. CARTER
Plenum, New York & London
1996

Transmission Electron Microscopy
L. REIMER
Springer Series in Optical Science, Springer, Berlin
1987

Technique

Formation simultanée de deux diagrammes de diffraction électronique
J. BEAUVILLAIN
Journal de Microscopie
1970, 4, 455-464

Application of a high voltage transmission scanning electron microscope
J.M. COWLEY, D.J. SMITH and G.A. SUSSEX
Proceedings 3rd Scanning Electron Microscope Symposium, Chicago
1970, 9, 16

Line patterns in wide angle convergent beam electron diffraction
D.J. SMITH and J.M. COWLEY
Journal of Applied Crystallography
1971, 4, 482-487

Etude de diagrammes de pseudo-lignes de Kikuchi obtenues par une méthode
d'observation en condenseur-objectif
J. BEAUVILLAIN et R. AYROLES
Physica Status Solidi A
1977, 44, 485

LACBED
M. TANAKA, R. SAITO, K. UENO and Y. HARADA
Journal of Electron Microscopy
1980, 29, 408-412

Zone-axis patterns formed by a new double-rocking technique
J.A. EADES
Ultramicroscopy
1980, 5, 71-74

Another way to form zone-axis patterns
J.A EADES
IOP Conference Series
1980, 52, 9-12

Development in CBED
J.W. STEEDS and R. VINCENT
Journal de Microscopie et de Spectroscopie Electroniques
1983, Vol. 8, 419-430

Zone axis diffraction patterns by the Tanaka method
J.A. EADES
Journal of Electron Microscopy Technique
1984, 1, 279-284

LACBED ZAP's
K.K. FUNG
Ultramicroscopy
1984, 12, 243-246

Simultaneous observation of bright field and dark field LACBED patterns
M. TERAUCHI and M. TANAKA
Journal of Electron Microscopy
1985, 34, 128-135

Simultaneous observation of zone axis patterns and $\pm$ g dark field patterns in CBED
M. TERAUCHI and M. TANAKA
Journal of Electron Microscopy
1985, 34, 347-356

Conventional TEM techniques in CBED
M. TANAKA
Journal of Electron Microscopy
1986, 35, 314-323

Convergent beam electron diffraction and imaging of strained layer superlattices
D.M. MAYER, H.L. FRASER, C.J. HUMPHREYS, R.V. KNOELL, R.D. FIELD,
J.B. WOODHOUSE and J.C. BEAN
Electron Microscopy and Analysis, IOP Conference Series
1986, 78, 49-52

Simultaneous observation of zone-axis pattern and $\pm$ g dark-field patterns in CBED
M. TERAUCHI and M. TANAKA
Proceedings ICEM-11, Kyoto
1986, 693-694

Simultaneous observation of a zone-axis pattern and $\pm$ g dark-field patterns in CBED
M. TERAUCHI and M. TANAKA

BIBLIOGRAPHY

JEOL News
1986, 24E,18-22

Convergent Beam Electron Diffraction
J.W STEEDS
In Microscopia Elettronica in Trasmissione e Techniche di Analisi di Superfici nella Scienza
dei Materiali. B, Microscopia Elettronica in Transmissione
Edizione ENEA, Roma
1986

A modified Goodman's technique for obtaining rocking-beam type patterns
Z. KANG
Journal of Electron Microscopy Technique
1986, 4, 343-346

Convergent beam diffraction
J.A. EADES and C.J. KIELY
IOP Conference Series
1987, 90, 109-114

Simultaneous observation of a zone axis and a $\pm g$ dark field pattern in CBED
M. TANAKA and M. TERAUCHI
JEOL News
1987, 25E, 2-6

Microdiffraction's contribution to microcharacterization
J.A. EADES
Ultramicroscopy
1988, 24, 143-154

Convergent-beam imaging - a transmission electron microscopy technique for investigating
small-localized distortions in crystals
C.J. HUMPHREYS, D.M. MAHER, H.L. FRAZER and D.J. EAGLESHAM
Philosophical Magazine
1988, 58, 787-798

CBED and CBIM from semiconductors and superconductors
C.J. HUMPHREYS, D.J. EAGLESHAM, D.M. MAHER and H.L. FRAZER
Ultramicroscopy,
1988, 26, 13-24

Skew thoughts on parallelism
K.K. CHRISTENSON and J.A. EADES
Ultramicroscopy
1988, 26, 113-132

Techniques of CBED
R. VINCENT
Journal of Electron Microscopy Technique
1989, 13, 40-50

Direct parallel detection of energy-resolved LACBED patterns
E. QUANDT, S. LA BARRE, H. NIEDRIG

Ultramicroscopy
1990, 33, 15-21

Identification of lattice defects by CBED
M. TANAKA, M. TERAUCHI and T. KANEYAMA
Journal of Electron Microscopy
1991, 40, 211-220

High-resolution shadow image superimposed on LACBED patterns: a method demonstrated
on Ge_xSi_{1-x} / Si
X.F. DUAN
Ultramicroscopy
1992, 41, 249-252

CBED
D. CHERNS
Journal de Physique IV
1993, 3, 2113-2122

DISLOCATIONS

CBED studies of crystal defects
D. CHERNS and A.R. PRESTON
Proceedings ICEM-11, Kyoto
1986, 721-722

Observation of Berry's geometrical phase in electron diffraction from a screw dislocation
D.M. BIRD and A.R. PRESTON
Physical Review Letters
1988, 61, 2863-2866

CBED studies of defects, strains and composition profiles in semiconductors
D. CHERNS
In Evaluation of Advanced Semiconducting Materials by Electron Microscopy,
D. Cherns Ed., NATO ASI Series B
1988, 59-74

Convergent beam diffraction studies of interfaces, defects and multilayers
D. CHERNS and A.R. PRESTON
Journal of Electron Microscopy Technique
1989, 13, 111-122

Measurement of Burger's vector from LACBED patterns
D.M. BIRD and A.R. PRESTON
IOP Conference Series
1989, 98, 123-126

Identification of lattice defects by CBED
M. TANAKA, and T. KANEYAMA
JEOL News
1989, 27E, 8-12

Distortion of the zeroth-order Laue zone pattern caused by dislocations in a silicon crystal
J. WEN, R. WANG and G. LU
Acta Crystallographica
1989, A45, 422-427

Simulation and application of the distorted ZOLZ patterns from dislocations in Si
G. LU, J.G. WEN, W. ZHANG and R. WAND
Acta Crystallographica A
1990, 46, 103-112

Applications of CBED techniques to the structural characterization of crystalline defects
R. PEREZ
Microbeam Analysis, San Francisco Press, San Francisco
1990, 354-363

Characterization of crystalline defects using CBED techniques
R. PEREZ
Proceedings ICEM-12, Seattle
1990, 520-521

CBED characterization of dislocations in GaS single crystals
C. DE BLASI, D. MANNO and A. RIZZO
Ultramicroscopy
1990, 33, 143-149

CBED analysis of extended defects in melt-grown GaSe single crystals
C. DE BLASI and D. MANNO
Proceedings ICEM-12, Seattle
1990, 494-495

Analysis of extended defects in melt-grown GaSe single crystal by CBED techniques
C. DE BLASI and D. MANNO
Ultramicroscopy
1991, 35, 71-76

Application of CBED and LACBED to the characterization of dislocations and stacking faults
J.P. MORNIROLI and J.W. STEEDS
IOP Conference Series
1991, 119, 417-420

Identification of lattice defects by CBED
M. TANAKA, M. TERAUCHI and T. KANEYAMA
Journal of Electron Microscopy
1991, 40, 211-220

Burgers vector determination of dislocations in an $Al_{70}Co_{15}Ni_{15}$ decagonal quasicrystal
Y. YA, R. WAN and J. FEN
Philosophical Magazine Letters
1992, 66, 197-201

Dislocation contrast in LACBED patterns
C.T. CHOU, A.R. PRESTON and J.W. STEEDS

Philosophical Magazine A
1992, 65, 863-888

CBED study of the Burgers vector of dislocations in an icosahedral Al-Cu-Fe alloy
M.X. DAI
Philosophical Magazine Letters
1992, 66, 235-240

The study of misfit dislocations in $In_xGa_{1-x}As$ / GaAs strained quantum well structures
J. WANG, J.W. STEEDS and D. A. WOOLF
Philosophical Magazine A
1992, 65, 829-839

Determining the Burgers vector of decorated dislocations in γ-TiAl by diffraction contrast and CBED
J.M.K. WIEZOREK, A.R PRESTON, S.A. COURT, H.L. FRASER and C.J. HUMPHREYS
Proceedings EUREM 92, Granada
1992, 1, 209-210

LACBED study of defects
Y. XIN and X.F. DUAN
Proceedings EUREM 92, Granada
1992, 1, 211-212

Study of imperfect crystals by CBED
D. CHERNS
Proceedings EUREM 92, Granada
1992, 1, 157-161

The Burgers vector of an edge dislocation in an $Al_{70}Co_{15}Ni_{15}$ decagonal quasicrystal
determined by means of CBED
Y. YA, and R. WAN
Journal of Physics
1993, 5, 195-200

Edge dislocations in icosahedral Al-Pd-Mn alloys
M.X. DAI
Philosophical Magazine A
1993, 67, 789-796

Characterization of dislocations in anisotropic materials by LACBED
J.P. MORNIROLI and P. VERMAUT
Journal de Physique IV
1993, 3, 2165-2168

A new six-dimensional Burgers vector of dislocations in icosahedral quasicrystals
determined by CBED
J. FENG, R. WANG and Z. WANG
Philosophical Magazine Letters
1993, 68, 321-326

Comment on CBED study of the Burgers vector of dislocations in icosahedral Al-Fe-Cu
alloy. Author's reply.

Philosophical Magazine Letters
1993, 68, 273-277

High-temperature deformation-induced defects and Burgers vector determination of dislocations in the $Al_{70}Co_{15}Ni_{15}$ decagonal quasicrystal
R. WANG, Y.F. YAN and K.H. KUO
Journal of Non-Crystalline Solids
1993, 153-54, 103-107

Characterization of the Burgers vectors of partial dislocations by LACBED
J.P. MORNIROLI and D. CHERNS
IOP Conference Series
1993, 138, 3, 153-156

Burgers vector of dislocations in an icosahedral $Al_{62}Cu_{25.5}Fe_{12.5}$ quasicrystal determined by means of CBED
R. WANG and M.X. DAI
Physical Review B
1993, 47, 15326-15329

Probing semiconductor interfaces by transmission electron microscopy
J.W. STEEDS and D. CHERNS
Philosophical Transactions of the Royal Society of London A
1993, 344, 545-556

CBED study of the Burgers vectors of dislocations in icosahedral quasicrystals
J. FENG and R. WANG
Philosophical Magazine A
1994, 69, 981-994

Fivefold and threefold-type dislocations in an Al-Cu-Fe icosahedral quasicrystal identified by CBED
J. FENG AND R. WANG
Philosophical Magazine Letters
1994, 69, 309-315

Reasonably good images of dislocations in LACBED patterns and effect of dislocation strain fields on the Bragg lines.
Y. XIN and X.F. DUAN
Ultramicroscopy
1994, 53, 159-165

Analysis of partial and stair-rod dislocations by LACBED
D. CHERNS and J.P. MORNIROLI
Ultramicroscopy
1994, 53, 167-180

Effects of dislocation strain fields on Bragg lines in an $Al_{70}Co_{15}Ni_{15}$ decagonal quasicrystal studied by an improved LACBED technique.
Y. YAN
Microscopy, Microanalysis, Microstructures
1994, 5, 183-187

Burgers vector determination of decorated dislocations in γ-TiAl by diffraction contrast and LACBED
J.M.K. WIEZOREK, A.R. PRESTON, S.A. COURT, H.L. FRASER and C.J. HUMPHREYS
Philosophical Magazine A
1994, 69, 285-299

Analysis of crystal defects by CBED techniques
D. CHERNS
Proceedings ICEM-13, Paris
1994, 893-894

Determination for core nature of partial dislocations in SiC
X.J. NING and P. PIROUZ
Proceedings ICEM-13, Paris
1994, 895-896

Characterization of the misfit nature of dislocations in boron-doped silicon
X.J. NING and P. PIROUZ
Proceedings ICEM-13, Paris
1994, 897-898

Diffraction contrast associated with concentric and pseudo-square loops in SiGe/Si
Y. ATICI and D. CHERNS
Proceedings ICEM-13, Paris
1994, 899-900

Characterization of crystal defects in quartz by LACBED
P. CORDIER, J.P. MORNIROLI and D. CHERNS
Philosophical Magazine A
1995, 72, 1421-1430

Study of dislocations generated by thermal cycling in Ni-Ti-Co shape memory alloys
J.L. PONS, L. JORDAN, J.P. MORNIROLI and R. PORTIER
Journal de Physique IV, Colloque C2
1995, 5, 293-298

Analysis of grain boundary dislocations by LACBED
J.P. MORNIROLI and D. CHERNS
Ultramicroscopy
1996, 62, 53-63

TEM investigation of dislocation microstructure of experimentally deformed silicate garnet
P. CORDIER, P. RATERRON and Y. WANG
Physics of the Earth and Planetary Interiors
1996, 97, 121-131

Analysis of dislocations by Convergent Beam Imaging (CBIM)
J.P. MORNIROLI and P. CORDIER
Proceedings EUREM-11, Dublin
1996, 2, 563-564

What is the difference between a dislocation crossing a bend contour and a bend contour crossing a dislocation?

BIBLIOGRAPHY

A.R. PRESTON and W.M. STOBBS
Proceedings EUREM-11, Dublin
1996, 2, 561-562

Computed and experimentally observed LACBED contrast from coherent precipitates and
decorated dislocations
J.M.K. WIEZOREK and H.L. FRASER
Proceedings EUREM-11, Dublin
1996, 2, 543-544

LACBED and CBIM analysis of dislocations in anisotropic materials
J.P. MORNIROLI, J. PONS and R. PORTIER
Proceedings EUREM-11, Dublin
1996, 2, 537-538

Analysis of dislocations and interfaces by LACBED
D. CHERNS
Proceedings EUREM-11, Dublin
1996, 2, 521-522

Characterization of dislocations in GaN by transmission electron diffraction and microscopy
techniques
F.A. PONCE, D. CHERNS, W.T. YOUNG and J.W. STEEDS
Applied Physics Letters
1996, 69, 770-772

Application of the Convergent Beam Imaging (CBIM) technique to the analysis of crystal
defects
J.P. MORNIROLI, P. CORDIER, E. VAN CAPPELLEN, J.M. ZUO and J.C.H. SPENCE
Microscopy, Microanalysis, Microstructures
1997, 8, 187-202

Observation of coreless dislocations in α-GaN
D. CHERNS, W.T. YOUNG, J.W. STEEDS, F.A. PONCE and S. NAKAMURA
Journal of Crystal Growth
1997, 178, 201-206

Characterisation of dislocations, nanopipes and inversion domains in GaN by transmission
electron microscopy
D. CHERNS, W.T. YOUNG and F.A. PONCE
Materials Science and Engineering
1997, B50, 76-81

Dislocations in 6H-SiC and their influence on electrical properties of n-type crystals
V. TILLAY, F. PAILLOUX, M.F. DENANOT, P.PIROUZ, J. RABIER, J.L. DEMENET and J.F.
BARBOT
European Physical Journal-Applied Physics
1998, 2, 111-115

Plastic deformation of silicate garnets. I - High-pressure experiments
V. VOEGELE, J.I. ANDO, P. CORDIER and R.C. LIEBERMANN
Physics of the Earth and Planetary Interiors
1998, 108, 305-318

Plastic deformation of silicate garnets II. Deformation microstructures in natural samples
V. VOEGELE, P. CORDIER, V. SAUTTER, T.G. SHARP, J.M. LARDEAUX, F.O MARQUES
Physics of the Earth and Planetary Interiors
1998, 108, 319-338

Large angle convergent beam electron diffraction determinations of dislocations Burgers
vector in synthetic stishovite
P. CORDIER and T.G. SHARP
Physics and Chemistry of Minerals
1998, 25, 548-555

TEM studies of defects in α-GaN
D. CHERNS
Proceedings ICEM-14, Cancun
1998, 385-386

Transmission electron microscopic studies of GaN grown on silicon carbide and sapphire by
laser induced molecular beam epitaxy
H. ZHOU, F. PHILLIPP, M. GROSS and H. SCHRODER
Materials Science and Engineering
1999, B68, 26-34

Elastic scattering of partially coherent beams of fast electrons by a crystal with a defect
N.I. BORGART
Acta Crystallographica A
1999, 55, 289-304

Dislocations in meteoritic and synthetic majorite garnets
V. VOEGELE, P. CORDIER, F. LANGENHORST and S. HEINEMANN
European Journal of Mineralogy
2000, 12, 696-702

Higher-order Laue zone line contrast in large-angle convergent-beam electron diffraction
around a dislocation
A. TODA, N. IKARASHI and H. ONO
Journal of Microscopy
2001, 239-245

STACKING FAULTS

Symmetry information from faulted layered crystals
W. JIANG and K.K. FUNG
Proceedings ICEM-12, Seattle
1986, 717-718

CBED studies of imperfect crystals
M. TANAKA and T. KANEYAMA
Proceedings ICEM-12, Seattle
1986, 203-206

High-energy electron diffraction from transverse stacking faults in the projection
approximation

D.E. JESSON and J.W. STEEDS
Ultramicroscopy
1989, 31, 399-430

Analysis of extended defects in melt-grown GaSe single crystal by CBED techniques
C. DE BLASI and D. MANNO
Ultramicroscopy
1991, 35, 71-76

CBED studies of imperfect gallium chalcogenides crystals
C. DE BLASI, D. MANNO and A. RIZZO
IOP Conference Series
1991, 119, 387-390

CBED and LACBED characterization of overlapping stacking faults and microtwins
J.P. MORNIROLI, A. LEFEBVRE, C.T. CHOU, G. LU and J.W. STEEDS
Proceedings EUREM-92, Granada
1992, 1, 1978-1980

Determination of the displacement vector of a stacking fault in TiO_2 by CBED
S. YAMADA and M. TANAKA
Journal of Electron Microscopy
1995, 44, 212-218

Determination of the orientation of a stacking fault by LACBED
X. WEI, X. DUAN and S.Q. WAN
Ultramicroscopy
1996, 66, 49-57

Structure of a stacking fault in the $(\bar{1}01)$ plane of TiO_2
S. YAMADA and M. TANAKA
Journal of Electron Microscopy
1997, 67-74

GRAIN BOUNDARIES - TWINS - INTERFACES

The symmetry of CBED from bicrystals containing a vertical grain boundary
F.W. SCHAPINK, S.K.E. FORGHANY and R.P. CARON
Philosophical Magazine A
1986, 53, 717-725

Determination of the interface atomic structure of a type $CoSi_2Si$ (111) using TEM techniques
M.A. AL KHAFAJI, D. CHERNS, C.J. ROSSOUW and R. HULL
IOP Conference Series
1991, 119, 51-54

{110} twinning in $YBa_2Cu_3O_{7-x}$
P.A. MIDGLEY, R. VINCENT and D. CHERNS
Philosophical Magazine A
1992, 66, 237-256

Characterization of grain boundary misorientation in α alumina by LACBED
J.P. MORNIROLI, A. LECLERE, W. SWIATNICKI and J.Y. LAVAL
Journal de Physique IV
1993, 3, 1455-1458

LACBED studies of grain boundaries
J.P. MORNIROLI, F. STRZELCZYK, A. REDJAIMIA and D. CHERNS
Proceedings ICEM-13, Paris
1994, 901-902

Analysis of grain boundaries by LACBED
J.P. MORNIROLI
Proceedings EUREM-11, Dublin
1996, 2, 529-530

Accurate measurement of grain boundary misorientation by LACBED
J.P. MORNIROLI
Interface Science
1997, 4, 273-283

LACBED study of a $\Sigma 3$ grain boundary in a Cu plus 6 at% Si alloy
H.S. KIM, P. GOODMAN, A. SCHWARTZMAN, P. TULLOCH and C.T. FORWOOD
Ultramicroscopy
1999, 77, 83-95

LOCAL STRAINS

Convergent beam diffraction studies of crystal defects
D. CHERNS and A.R. PRESTON
Proceedings ICEM-12, Seattle
1986, 721-722

Measurement of strain in silver halide particles by CBED
R. VINCENT, A. R. PRESTON and M.A. KING
Ultramicroscopy
1988, 24, 409-420

Electron diffraction studies of strain in epitaxial bicrystals and multilayers
D. CHERNS, C.J. KIELY and A.R. PRESTON
Ultramicroscopy
1988, 24, 355-370

A study of lattice distortion in Ge_xSi_{1-x} superlattice by CBED
X. DUAN, K.K. FUNG and Y.M. CHU
Proceedings ICEM-12, Seattle
1990, 614-615

CBED studies of strain in Si/SiGe superlattices
D. CHERNS, R. TOUAITIA, A.R. PRESTON and C.J. ROSSOUW
Philosophical Magazine A
1991, 64, 597-612

BIBLIOGRAPHY

A treatment of dynamical diffraction for multiply layered structures
C.J. ROSSOUW, M. AL-KHAFAJI, D. CHERNS, J.W STEEDS and R. TOUAITIA
Ultramicroscopy
1991, 35, 229-236

CBED study of the $Ge_{0.5}Si_{0.5}$ / Si strained-layer superlattices grown by molecular beam epitaxy
X.F. DUAN, K.K. FUNG, Y.M. CHU, C. CHENG and G. L. ZHOU
Philosophical Magazine Letters
1991, 63, 79-85

Investigation of strain in $Si_{1-x}Ge_x$ / Si heterostructures and local isolation structures by CBED
A. ARMIGLIATO, R. BALBONI, S. FRABBONI and J. VANHELLEMONT
Solid State Phenomena
1993, 32-33, 547-552

Application of convergent beam illumination methods to the study of lattice distortion across the interface
Z. LILIENTAL-WEBER, T. KANEYAMA, M. TERAUCHI and M. TANAKA
Journal of Crystal Growth
1993, 132, 103-114

Strains in crystals with amorphous surface-films studied by CBED and high-resolution imaging
F. BANHART
Ultramicroscopy
1994, 56, 233-240

Preliminary studies of the interface strain field in an Al / Al_2O_3 composite by means of CBED
Y. YAN, R. WANG, H. ZOU and M.X DAI
Scripta Metallurgica et Materialia
1994, 30, 885-886

Effect of elastic relaxation on LACBED from cross-sectional specimens of $Ge_xSi_{(1-x)}$ / Si strained-layer superlattice
X.F. DUAN, D. CHERNS and J.W. STEEDS
Philosophical Magazine A
1994, 70, 1091-1105

Detection of local lattice strains by CBED
Y. TOMOKIYO
Proceedings of the NIRIM International Symposium on Advanced Materials '94
1994, 40-45

Surface relaxation of strained heterostructures revealed by Bragg line splitting in LACBED patterns
C.T. CHOU, S.C. ANDERSON, D.J.H. COCKAINE, A.Z. SIKORSKI and M.R. VAUGHAN
Ultramicroscopy
1994, 55, 334-347

Estimation of strain state in cross-sectional specimens of Ge_xSi_{1-x} / Si strained-layer superlattices by LACBED
X.F. DUAN, D. CHERNS and J.W. STEEDS

Proceedings ICEM-13, Paris
1994, 903-904

Application of surface relaxation effects to the measurement of built in strain in semiconductors heterostructures
C.T. CHOU, S.C. ANDERSON, D.J.H. COCKAINE, A.Z. SIKORSKI and M.R. VAUGHAN
Proceedings ICEM-13, Paris
1994, 905-906

A new application of LACBED to measure the strain in strained layers
Y. ATICI and D. CHERNS
Proceedings. ICEM-13, Paris
1994, 907-908

Study of strain distribution in the active region of MOS structure using a LACBED technique
Y. OSHIMA
Proceedings ICEM-13, Paris
1994, 909-910

LACBED studies of lattice bendings in epitaxial Si / Sio$_2$ systems
F. BANHART, F. PHILLIP, N. NAGEL, I. SILIER and E. CECH
Proceedings ICEM-13, Paris
1994, 911-912

Determination of lattice strains near oxygen precipitates in Czochralski Si by CBED
J. NAKASHIMA, Y. TOMOKIYO, T. OKUYAMA, S. SADAMITSU, S. SUMITA and T. SHIGEMATSU
Proceedings ICEM-13, Paris
1994, 913-914

Analysis of local lattice strains around plate-like oxygen precipitates in Czochralski-silicon wafers by convergent-beam electron diffraction
T. OKUYAMA, N. NAKAYAMA, S. SADAMITSU, J. NAKASHIMLA and Y. TOMOKIYO
Japanese Journal of Applied Physics, Part 1
1997, 36, 3359-3365

Determination of residual strains with CBED / LACBED techniques
F. WU and J. ZHU
Science in China, Series E
1998, 41, 121-129

Profiling Ge islands in Si by large angle convergent beam electron diffraction
D. CHERNS, A. HOVSEPIAN and W. JAEGER
Journal of Electron Microscopy
1998, 47, 211-215

Determination of interfacial residual stress field in an Al-Al$_2$O$_3$ composite using convergent-beam electron diffraction technique
H.M. ZOU, J. LIU, D.H. DING, R.H. WANG, L. FROYEN and L. DELAEY
Ultramicroscopy
1998, 72, 1-15

Strain and Ge concentration determinations in SiGe/Si multiple quantum wells by TEM methods
A. BENEDETTI, D.J. NORRIS, C.D.J. HETHERINGTON, A.C. CULLIS, A. ARMIGLIATO, R. BALBONI, D.J. ROBBINS and D.J. WALLIS
Proceedings Microscopy of Semiconducting Materials 1999
IOP Conference Series
1999, 164, 212-222

Microstructure evolution accompanying high temperature; uniaxial tensile creep of self-reinforced silicon nitride ceramics
Q. WEI, J. SANKAR, A.D. KELKAR and J. NARAYAN
Materials Science and Engineering A
1999, 272, 380-388

Investigation of strain distribution in LOCOS structures by dynamical simulation of LACBED patterns
F. WU, A. ARMIGLIATO, R. BALBONI and S. FRABBONI
Ultramicroscopy
1999, 80, 193-201

Epitaxial stress study by large angle convergent beam electron diffraction and high-resolution transmission electron microscopy moiré fringe patterns
F. PAILLOUX, R. J. GABORIAUD, C. CHAMPEAUX and A. CATHERINOT
Materials Science and Engineering A
2000, 288, 244-247

Stress relaxation in c(perpendicular to)-c(parallel to) YBaCuO thin films on MgO substrate studied by LACBED
F. PAILLOUX and R. J. GABORIAUD
Thin Solid Films
2000, 368, 142-146

Electron diffraction (LACBED) and HRTEM moiré fringe patterns study of stress in YBaCuO thin film on MgO
F. PAILLOUX, R. J. GABORIAUD
Journal de Physique IV
2000, 10, 131-135

Dynamical simulation of LACBED patterns in cross-sectioned heterostructures
F. WU, A. ARMIGLIATO, R. BALBONI and S. FRABBONI
Micron
2000, 31, 211-216

Crystalline growth rate and microstructure in YBaCuO thin films
F. PAILLOUX, R. J. GABORIAUD, C. CHAMPEAUX and A. CATHERINOT
Physica C
2001, 351, 9-12

AFM, SEM, EDX and HRTEM study of the crystalline growth rate anisotropy-induced internal stress and surface roughness of YBaCuO thin film
F. PAILLOUX, R. J. GABORIAUD, C. CHAMPEAUX and A. CATHERINOT
Materials Characterization
2001, 46, 55-63

A study of interfacial residual stress field in a K2O centre dot 6TiO(2w) / Al composite by
LACBED and 3-D finite element method
X.B. LI, H.M. ZOU and J. PAN
Materials Science and Engineering A
2001, 308, 65-73

LAYERS AND MULTILAYERS

Dynamical diffraction from modulated structures
D.M. BIRD and R.L. WITHERS
IOP Conference Series.
1985, 78, 31-34

Structure of AlGaAs / GaAs multilayers imaged in superlattice reflections
R. VINCENT, D. CHERNS, S.J. BAILEY and H. MORKOC
Philosophical Magazine Letters
1987, 56, 1-6

Application of WACBED to superlattice structures in semiconductors
S. McKERNAN
Proceedings EMSA, San Francisco Press
1987, 45, 44-45

Analysis of LACBED patterns from InP / InGaAs multiple quantum well sample by dynamical
theory
I.K. JORDAN, A.R. PRESTON, L.C. QIN and J.W. STEEDS
IOP Conference Series
1989, 98, 3, 131-134

Profiling semiconductor multilayers structures by CBED and CBIM techniques
D. CHERNS
Proceedings ICEM-12, Seattle
1990, 612-613

A treatment of dynamical diffraction for multiply layered structures
C.J. ROSSOUW, M. AL-KHAFAJI, D. CHERNS, J.W. STEEDS and R. TOUAITIA
Ultramicroscopy
1991, 35, 229-236

CBED studies of modulations in semiconductor superlattices
K.K. FUNG, Q.H. XIE and X.F. DUAN
Ultramicroscopy
1991, 38, 143-148

High resolution shadow image superimposed on LACBED patterns: a method demonstrated
in Ge_xSi_{1-x}/Si superlattices
X.F. DUAN
Ultramicroscopy
1992, 41, 1-3, 249-252

LACBED study of the Ge_xSi_{1-x} / Si strained layer superlattices
X.F. DUAN

Proceedings EUREM-92, Granada
1992, 1, 183-184

Electron microscopy of ultra-thin single quantum wells
N. GRIGORIEF, M.J. YATES, D. CHERNS, M.HOCKLY, S.D. PERRIN and M.R. AYLETT
Proceedings EUREM 92, Granada
1992, 1, 121-122

Electron microscopy of ultra-thin buried layers in InP and InGaAs
N. GRIGORIEV, D. CHERNS, M.J. YATES, M. HOOKLY, S.D. PERRIN and M.R. AYLETT
Philosophical Magazine A
1993, 68, 121-122

Dynamical LACBED analysis of Si / SiGe and Si / SiB multilayer structures
C.J. ROSSOUW and D.D. PEROVIC
Ultramicroscopy
1993, 48, 49-61

Observation of crystal distortion in SiGe / Si superlattice using a new application of LACBED
Y. ATICI and D. CHERNS
Ultramicroscopy
1995, 58, 435-440

Effect of static disorder on LACBED patterns of $Si_{1-x}Ge_x$ / Si heterostructures
S. FRABBONI, F. GAMBETTA, R. BALBONI, A. ARMIGLIATO and E. VAN CAPPELLEN
Proceedings EUREM-11, Dublin
1996, 2, 531

Investigation of SiGe quantum dot structures by large angle CBED and finite element analysis
A. HOVSEPIAN, D. CHERNS and W. JAEGER
IOP Conference Series
1997, 153, 413-416

Analysis of ultra-thin Ge layers in Si by large angle convergent beam electron diffraction
A. HOVSEPIAN, D. CHERNS and W. JAEGER
Philosophical Magazine A
1999, 79, 1395-1410

LACBED measurement of the chemical composition of a thin $In_xGa_{1-x}As$ layer buried in a GaAs matrix
D. JACOB, Y. ANDROUSSI and A. LEFEBVRE
Ultramicroscopy
2001, 89, 299-303

STRUCTURES AND STRUCTURE FACTOR

Crystallographic data from zone axis patterns
J.W. STEEDS, J.R. BAKER and R. VINCENT
Proceedings ICEM-10, Hamburg
1982, 617-624

Structure of AuGeAs determined by CBED. I - Derivation of basic structure. II - Refinement of structural parameters
R. VINCENT, D.M. BIRD and J.W. STEEDS
Philosophical Magazine A
1984, 50, 745-763, 765-786

CBED from organic crystals
R. VINCENT
IOP Conference Series
1985, 78, 427-428

Phase analysis in the Ni-Ge-P system by electron diffraction
R. VINCENT and S.F. PRETTY
Philosophical Magazine A
1986, 53, 843-862

Measurement of kinetic intensities from LACBED patterns
R. VINCENT and D.M. BIRD
Philosophical Magazine Letters
1986, 53A, L35-L40

Quasi two-beam description of the systematic row Kossel pattern. Application to determination of low order structure factors in GaAs and $Ga_{1-x}Al_xAs$
K. GJØNNES and J.GJØNNES
IOP Conference Series.
1988, 93, 41-42

Structural modifications induced by iodine in melt grown GaSe
C. DE BLASI, A.M. MANCINI, D. MANNO and A. RIZZO
IOP Conference Series
1988, 93, 123-124

Structural transformations in $YBa_2Cu_3O_{7-x}$
P.A. MIDGLEY, R. VINCENT and D. CHERNS
IOP Conference Series
1989, 98, 499-502

Analysis by electron diffraction of rhombohedral Al-Ge phase
R. VINCENT and D. EXELBY
Proceedings ICEM-12, Seattle
1990, 524-525

Determination of the static displacement of atoms by means of LACBED
Y. TOMOKIYO and T. KUROIWA
Proceedings ICEM-12, Seattle
1990, 526-527

New phases in the La-Cu-O system
P.A. MIDGLEY and R. VINCENT
IOP Conference Series
1991, 119, 413-416

BIBLIOGRAPHY

Crystallographic study of lanthanum aluminate by CBED
C.Y. YANG, Z.R. HUANG, W.H. YANG, Y.Q. ZHOU and K.K. FUNG
Acta Crystallographica
1991, A47, 703-706

Applications of CBED to extract quantitative information in materials science
Y. TOMOKIYO
Journal of Electron Microscopy
1992, 41, 403-413

Symmetry determination and Pb-site ordering analysis for the
n=1,2, $Pb_xBi_{2-x}Sr_2Ca_{n-1}Cu_nO_{4+2n+\delta}$ compounds by convergent-beam and selected-area
Electron Diffraction
P. GOODMAN, P. MILLER, T.J. WHITE and R.L. WITHERS
Acta Crystallographica
1992, B48, 376-389

Reassessment of the symmetry of the 221 PbBiSrCaCuO structure using LACBED and
high-resolution SAD: the relevance of Cowley's theory of disorder scattering to a real-space
structural analysis
P. GOODMAN and P. MILLER
Ultramicroscopy
1993, 52, 549-556

Electron diffraction measurements on chromium oxide
C.J. ROSSOUW and J.M. HAMPIKIAN
Philosophical Magazine A
1993, 67, 849-863

Structure of metastable Al-Ge phase determined from LACBED patterns
R. VINCENT and D.R. EXELBY
Philosophical Magazine B
1993, 68, 513-528

Debye-Waller factor determination from LACBED patterns
A.R. PRESTON, W.G. BURGESS, C.J. PICKUP and C.J. HUMPHREYS
IOP Conference Series
1993, 138, 145-148

Dynamical effects of high-energy electron diffraction observed in ZnS
H. MATSUHUTA and G. GJØNNES
IOP Conference Series
1993, 138, 149-152

A study of the structure factors in rutile-type SnO_2 by high energy electron diffraction
H. MATSUHUTA,, J. GJØNNES and J. TAFTO
Acta Crystallographica A
1994, 50, 115-123

Bloch-wave degeneracies and non-systematic critical voltage: a method for structure factor
determination
H. MATSUHUTA and G. GJØNNES

Acta Crystallographica A
1994, 50, 107-115

LACBED determination of structure factors and alloy composition of GeSi / Si SLC
S.Q. WANG and L.M. PENG
Ultramicroscopy
1994, 55, 67-73

Determination of structure factors, lattice strains and accelerating voltage by energy-filtered CBED
C. DEININGER, G. NECKER and J. MAYER
Ultramicroscopy
1994, 54, 15-30

Structure refinement from CBED integrated intensities
K. GJØNNES and J. GJØNNES
Proceedings ICEM-13, Paris
1994, 853-854

Identification of forbidden reflections by LACBED
J.P. MORNIROLI, M. MANGOLD and O. RICHARD
Proceedings ICEM-13, Paris
1994, 915-916

Determination of structure factors of Al_2O_3 by means of LACBED
TOMOKIYO, YOSHITUGU, KIOROIWA, TAKEHARU, HAYASHI and YASUNORI
Materials Transactions, Journal of the Institute of Metals
1995, 36,1344-1348

Effect of Mn doping on charge density in γ TiAl by quantitative CBED
R. HOLMESTAD, J.M. ZUO, J.C.H. SPENCE, R. HOIER and Z. HORITA
Philosophical Magazine A
1995, 72, 579-601

Quantitative electron microdiffraction
J.C.H. SPENCE
Journal of Electron Microscopy
1996, 45, 19-26

CBED study of polytypism in chlorine doped GaSe grown from the melt
D. MANNO, G. MICCOLI, A. SERRA and A. TEPORE
Proceedings EUREM-11, Dublin
1996, 2, 539

Application of the large angle convergent beam electron diffraction technique to the characterization of martensitic phases in Cu-Zn-Al alloys
A. TOLLEY and A. CONDO
Materials Science and Engineering A
1999, 275, 347-351

Quantitative electron diffraction study of cation configuration and irradiation induced displacement in magnesium aluminate spinel crystal
T. SOEDA, S. MATSUMURA, J. HAYATA and C. KINOSHITA

Journal of Electron Microscopy
1999, 48, 531-536

Effects of static disorder on LACBED patterns of single crystal silicon implanted with hydrogen
S. FRABBONI, F. GAMBETTA, R. TONINI, R. BALBONI and A. ARMIGLIATO
IOP Conference Series
1997, 157, 415-418

GLIDE PLANES AND SCREW AXES

Glide planes and screw axes in CBED: the standard procedure
J.A. EADES
Microbeam Analysis, San Francisco Press
1988 75-80

HOLZ lines in dark field convergent beam discs
J.A. EADES and R. JAMES
IOP Conference Series
1991, 119, 361-365

CRITICAL VOLTAGE

Observation of accidental Bloch wave degeneracies at zone-axis critical voltages in high-energy electron diffraction
H. MATSUHATA and J.W. STEEDS
Philosophical Magazine B
1987, 55, 39-54

Zone-axis critical voltages and LACBED at 200-400 kV
J.W. STEEDS
Ultramicroscopy
1988, 26, 1-12

Bloch-wave degeneracies and non-systematic critical voltage: a method for structure factor determination
H. MATSUHATA and J. GJØNNES
Acta Crystallographica A
1994, 50, 107-115

LACBED determination of structure factors and alloys composition of GeSi / Si SLS
SQ WAN and LM PENG
Ultramicroscopy
1994, 55, 67-73

A study of structure factors in rutile-type SnO_2 by high-energy electron
H. MATSUHATA and J. GJØNNES
Acta Crystallographica A
1994, 50, 115-123

COHERENT DIFFRACTION

Coherent overlapping LACBED patterns in 6H SiC
R. VINCENT, P.A. MIDGLEY, P. SPELLWARD and J.W. STEEDS
Ultramicroscopy
1993, 50, 365-376

Observation of lattice fringes in CBED
M. TERAUCHI, K. TSUDA, O. KAMIMURA, M. TANAKA, T. KANEYAMA and T. HONDA
Ultramicroscopy
1994, 54, 268-275

Observation of coherent CBED patterns from small lattice spacings and stacking faults
K. TSUDA, M. TERAUCHI, M. TANAKA, T. KANEYAMA and T. HONDA
Proceedings ICEM-13, Paris
1994, 865

Observation of interference fringes with a lattice spacing of 1.92 Å in CBED
K. TSUDA, M. TERAUCHI, M. TANAKA, T. KANEYAMA and T. HONDA
Journal of Electron Microscopy
1994, 43, 173-175

Interferometry by coherent convergent-beam electron diffraction
K. TSUDA and M. TANAKA
Journal of Electron Microscopy
1996, 45, 59-63

SIMULATION

Computer simulation of zone-axis patterns
D.E. JESSON
Proceedings Electron Microscopy Society of Southern Africa
1985, 15,15-16

EMS - A software package for electron diffraction analysis and HREM simulation in materials science
P.A. STADELMANN
Ultramicroscopy
1987, 21, 131-146

Quantitative absorption corrections for electron diffraction: correlation between theory and experiment.
CJ. ROSSOUW, P.R. MILLER, J. DRENNAN and L.J. ALLEN
Ultramicroscopy
1990, 34, 149-163

LACBED with quantitative anomalous absorption corrections
C.J. ROSSOUW, P.R. MILLER and L.J. ALLEN
Proceedings Microbeam Analysis 1990, San Francisco Press
1990, 337-340

LACBED with quantitative anomalous absorption corrections
C.J. ROSSOUW, L.J. ALLEN and P.R. MILLER
Proceedings ICEM-12, Seattle
1990, 28, 528-529

Many-beam simulations and observations of LACBED imaging of crystal defects
S.Q. WANG, L.M. PENG, Y. XIN, Y.M. CHU and X.F. DUAN
Philosophical Magazine Letters
1992, 66, 225-233

Experimental observation and computer simulation of HOLZ patterns of Al-Co-Ni decagonal quasicrystals
Y. YA, R. WAN, J. GU, M. DA and H. LUNXIONG
Philosophical Magazine Letters
1992, 65, 33-41

Matrix description of dynamical HOLZ diffraction tested on the strained layer superlattice Si/GeSi.
S.Q. WANG, L.M. PENG, X.F. DUAN and Y.M. CHU
Ultramicroscopy
1992, 45, 405-409

Quantitative TDS absorption corrections for LACBED patterns from GaP and InAs
L.J. ALLEN, C.J. ROSSOUW and A.G. WRIGHT
Ultramicroscopy
1992, 40, 109-119

Computation and measurement of characteristic energy-loss LACBED patterns of molybdenum selenide
A. WEICKENMEIER, E. QUANDT, H.KOHL, H. ROSE and H. NIEDRIG
Ultramicroscopy
1993, 49, 210-219

Smooth parabolas in transmission electron diffraction patterns
R. JAMES, D.M. BIRD and A.G. WRIGHT
Acta Crystallographica A
1994, 50, 357-366

LACBED on partial dislocation dipoles in TiAl: experimental observations and simulations
B. VIGUIER and R. SCHAÜBBLIN
Proceedings EUREM-11, Dublin
1996, 2, 557

X-Ray multislice computation using the Moodie-Wagenfeld equations: divergent-beam pattern simulation in three-beam and six-beam Laue cases
P. GOODMAN and L. LIU
Acta Crystallographica, A
1999, 55, 246-257

ENERGY FILTERING

Measurement and calculations of Bloch-waves localization effects in graphite and lithium fluoride
E. QUANDT, H.J. KOHL, H. NIEDRIG and D. RUBESAME
IOP Conference Series, 1988, 93, 51-52

Direct parallel detection of energy-resolved LACBED patterns
E. QUANDT, S. LA BARRE and H. NIEDRIG
Ultramicroscopy, 1990, 33, 15-21

Investigation of energy spectroscopic intensity distribution in LACBED patterns of $MoSe_2$
E. QUANDT, A. WEIKENMEIER, H. KOHL and H. NIEDRIG
Proceedings ICEM-12, Seattle, 1990, 44-45

Effect of energy filtering in LACBED patterns
I.K. JORDAN, C.J. ROSSOUW and R. VINCENT
Ultramicroscopy, 1991, 35, 237-243

Omega energy filtered CBED
C. DEININGER and J. MAYER
Proceedings EUREM-92, Granada, 1992, 1, 181-182

Computation and measurement of characteristic energy-loss LACBED patterns of molybdenum selenide
A. WEIKENMEIER, E. QUANDT, H. KOHL, H. ROSE and H. NIEDRIG
Ultramicroscopy, 1993, 49, 210-219

Benefits of energy filtering for advanced CBED patterns
W.G. BURGESS, A.R. PRESTON, G.A. BOTTON, N.J. ZALUZEC and C.J. HUMPHREYS
Ultramicroscopy, 1994, 55, 276-283

Energy-filtered imaging and diffraction: a prologue to the special section of the Electron Optics Bulletin
M. OTTEN
Philips Electron Optics Bulletin, 1994, 133, 22-24

Quantitative electron diffraction and plasmon imaging with a post-column imaging filter
G. BOTTON, W.G. BURGESS, S.L. CULLEN and C.J. HUMPHREYS
Philips Electron Optics Bulletin, 1994, 133, 45-48

Incoherent contrast under dynamical diffraction conditions
C.J. ROSSOUW
Ultramicroscopy, 1995, 58, 211-222

Electron Spectroscopic Diffraction
J. MAYER, C. DENINGER and L. REIMER
In Energy-Filtering Transmission Electron Diffraction. L. REIMER, Ed.
Springer Series in Optical Sciences, Springer-Verlag Berlin Heidelberg
1995, 71, 291-345

INDEX

INDEX

Saint-Paul Imprimeur, 55000 Bar le Duc – Dépôt légal : juin 2002 – N° 06-02-0745

Milton Keynes UK
Ingram Content Group UK Ltd.
UKHW022054141024
449569UK00031B/1632